ZEITSCHRIFT FÜR ALLGEMEINE MIKRO- -BIOLOGIE

ORPHOLOGIE, PHYSIOLOGIE,

ENETIK UND ÖKOLOGIE DER

MIKROORGANISMEN

HEFT 7 · 1977 · BAND 17

KADEMIE-VERLAG · BERLIN

20,— M

34112

INHALTSVERZEICHNIS HEFT 7

Kurze Originalmitteilung

CONTENTS OF NUMBER 7

Short Note

Dokumentationsfahne
zu Heft XVII/7 der „Allgemeinen Mikrobiologie"

M. Hasan and J. B. Hall: Dissimilatory nitrate reduction in *Clostridium tertium*. Z. Allg. Mikrobiol., 17, 501–506 (1977).

Fermentation balance studies were carried out on *Clostridium tertium* grown with and without nitrate in the medium. Nitrate reduction increased the efficiency of energy production from glucose by permitting the utilization of additional sites of substrate level phosphorylation. The effect was even more dramatic in *C. tertium* than in *C. perfringens*, with increased cell yields of about 30% being observed in the former compared with 20% in the latter. Unlike *C. perfringens*, *C. tertium* responded to the presence of nitrate in the medium with an increased growth rate. A slight increase in the Y_{ATP} of these cultures was also observed and quantitatively, this appeared to be consistent with the prediction of STOUTHAMER and BETTENHAUSSEN that Y_{ATP} will vary with the growth rate. Thus, *C. tertium*, like *C. perfringens*, was able to use nitrate as an electron acceptor in conjunction with its energy metabolism, suggesting that this trait may be widespread among the nitrate-reducing anaerobes.

M. Iida and H. Iizuka: Introduction of a 16α-hydroxyl function into estrone by *Streptomyces roseochromogenes*. Z. Allg. Mikrobiol., 17, 507–512 (1977).

S. roseochromogenes (NRRL B-1233) converted estrone predominantly into its 16α-hydroxyl derivative. Chemical and spectroscopic (UV, IR, NMR, MS) methods were used in establishing the structure and stereochemistry of the oxidation product. The product was assigned as the structure of 16α-hydroxyestrone (yield, 17%). No other oxidation product was detected in this experiment. An interrelationship between cell growth and 16α-hydroxy-estrone formation was observed. Also, 16α-hydroxylation of estrone was observed in resting cell experiment. 16α-Hydroxylase showed good activity at 3.7×10^{-4} M of estrone concentration and was completely inhibited by 1.1×10^{-3} M. These results indicate the presence of a constitutive 16α-hydroxylase in the organism investigated.

J. K. Ladha and H. D. Kumar: Some characteristics of two morphological mutants of *Nostoc linckia* induced by nitrosoguanidine. Z. Allg. Mikrobiol., 17, 513–519 (1977).

The blue-green alga *N. linckia* was treated with nitrosoguanidine and two classes of morphological mutant clones were isolated. One class shows certain abnormal phenotypic features of vegetative cells, spores, and heterocysts. It has increased heterocyst frequency and impaired growth rate. The other class exhibits an altered heterocyst spacing pattern. Both classes of mutants have reduced nitrogenase activity.

Dokumentationsfahne zu Heft XVII/7 der „Allgemeinen Mikrobiologie“

H. **Musielski**, W. **Mann**, B. **Degen**, S. **Michel und** G. **Richter: Einfluß der Temperatur auf die Transkription der T3-DNA durch T3-spezifische RNA-Polymerase in zellfreien Extrakten von *Escherichia coli* CRT266.** Z. Allg. Mikrobiol., **17**, 521–529 (1977).

In zellfreien Extrakten aus *E. coli* CRT266, die 9 min nach Infektion mit T3-Phagen hergestellt worden waren, wird RNA fast ausschließlich durch die T3-induzierte RNA-Polymerase synthetisiert. Die freie T3-RNA-Polymerase (nicht DNA-gebundenes Enzym) wird bei 42 °C schnell inaktiviert. Bei dieser Temperatur sinkt ihre Aktivität innerhalb von 10 min um 90%. In Gegenwart von GTP, ATP, UTP und T3-DNA, wird ein stabiler Initiationskomplex aus Enzym und T3-DNA, der wesentlich stabiler gegen Hitzeinaktivierung ist, ausgebildet. Nach 10 min bei 42 °C beträgt unter diesen Bedingungen der Verlust an Aktivität 10%. Untersuchungen über den zeitlichen Ablauf der RNA-Synthese zeigen, daß bei 37 °C und 42 °C eine Reinitiation möglich ist. Hingegen kommt es bei 44 °C nach 3 min zu einem abrupten Abbruch der RNA-Synthese, was zeigt, daß bei dieser Temperatur keine Reinitiation mehr stattfindet. Die Elongation schon initiierter T3-RNA-Ketten ist wesentlich resistenter gegen Hitzeinaktivierung und verläuft bei 44 °C mit derselben Geschwindigkeit wie bei 37 °C und 42 °C.

T. A. **Petrova,** W. A. **Knorre,** R. **Guthke** and F. **Bergter: A mathematical model of microbial growth including an intermediate. I. Growth on bach cultures.** Z. Allg. Mikrobiol., **17**, 531–542 (1977).

A mathematical model of microbial growth is presented and examined which, in contrast to the well-known Monod model, includes transitions from one cell "bottle-neck" to another. This is achieved by introducing an intermediate product in the model. Three variants of the model for different regulatory functions of the intermediate are considered. The results permit to describe a set of experimentally observable microbial growth curves. According to the model, the shape of the growth curves, the kinetics of substrate consumption, and changes of intermediate concentration depend on culture prehistory and the nature of the intermediate regulatory function.

G. **Reuter,** H. **Köster** und B. **Liebermann: Physiologie und Biochemie der Streptomyceten. XIII. Biosynthese von Paromomycin unter Einsatz von ^{14}C-Glucose, -Glucosamin, -2-Desoxystreptamin und -Ribose durch *Streptomyces albus var. metamycinus nov. var.*** Z. Allg. Mikrobiol., **17**, 543–547 (1977).

The distribution of radioactivity in paromomycin ascertained after application of ^{14}C-D-glucose, ^{14}C-D-glucosamine, ^{14}C-2-deoxystreptamine, respectively, ^{14}C-D-ribose is taken as a basis for a biosynthesis scheme: While ribose bound in the antibiotic originates from glucose by oxidation and following decarboxylation, glucosamine is formed via fructose-6-phosphate. Paromose I arises from glucosamine, but not the cyclohexan derivative 2-deoxystreptamine, whose biosynthesis pathway is directly branching off glucose.

Dokumentationsfahne
zu Heft XVII/7 der „Allgemeinen Mikrobiologie“

D. **Schröder**, Hiltraut **Mach** und F. **Mach: DNS-Stoffwechsel einer temperatursensitiven filamentösen Mutante von *Bacillus subtilis* SB 19.** Z. Allg. Mikrobiol., **17**, 549–558 (1977).

Die temperatursensitive filamentöse Mutante *ts* 33-6 von *B. subtilis* SB 19 zeichnet sich durch unterschiedliches Teilungsvermögen bei restriktiver und permissiver Temperatur aus (30 °C Kurzstäbchen, 46 °C Filamente). Sie verfügt unter restriktiven Verhältnissen (46 °C) über einen funktionsfähigen Replikationsmechanismus. Die während der Sporenkeimung bei 45 °C begonnene DNS-Replikation kann auch unter restriktiven Temperaturbedingungen beendet werden. Kurzzeitige Beeinflussung gekeimter Sporen durch Inkubation bei 48 °C und anschließender shift down zu 46 °C verändern den DNS-Syntheseablauf, verhindern jedoch die Filamentbildung nicht. Teilungsinduktion durch Temperaturwechsel von 46 °C zu 30 °C ist mit gesteigerter DNS-Synthese gekoppelt. Gegenüber einer bei 46 °C verbleibenden Kontrolle verstärkt sich die DNS-Syntheseaktivität um *ca.* 130%. Die DNS-Synthese nach Temperaturerniedrigung läuft unabhängig von der Proteinsynthesehemmung durch Chloramphenicol ab. Auch bei blockierter RNS-Synthese in aminosäurefreiem Medium ist die DNS-Replikation möglich. Lediglich Mitomycin C in geeigneten Konzentrationen hemmt bzw. beeinflußt die DNS-Replikation nach shift down.

G. **Strohbach** und Sigrid **Kretschmer: Einbau von Thymidin in die DNS von Actinomyceten. I. Der Einbau exogenen Thymidins in die DNS von *Thermoactinomyces vulgaris*.** Z. Allg. Mikrobiol., **17**, 559–568 (1977).

The incorporation of exogenous thymidine and thymine into acid-insoluble material of *T. vulgaris* has been studied during germination and subsequent growth. Thymine is not incorporated. The incorporation of thymidine stops after a short time due to the rapid breakdown of thymidine to thymine and deoxyribose-1-phosphate by the inducible thymidine phosphorylase. Deoxyadenosine enhances the incorporation of thymidine as well as of thymine and prolongs the time of uptake. Uridine stimulates only the incorporation of thymidine but not of thymine. These effects can be explained by the function of these substances within the salvage pathway. Deoxyadenosine acts as donor of deoxyribosyl groups being necessary for the conversion of thymine to thymidine by thymidine phosphorylase and uridine inhibits thymidine phosphorylase, and thereby it prevents the degradation of thymidine to thymine. Thymidine is incorporated into alkali, RNase-, and protease-stable hot TCA-soluble and DNase-sensitive material. That means that the cellular DNA of *T. vulgaris* can be specifically labelled by radioactive thymidine in the presence of deoxyadenosine and uridine, respectively.

Dokumentationsfahne
zu Heft XVII/7 der „Allgemeinen Mikrobiologie"

H.-E. **Jacob**, W. **Fleck** und D. **Strauss: Photodynamische Wirkung und Reparatur-Inhibition photodynamisch induzierter DNA Schäden in *Proteus mirabilis* durch das Anthracyclin-Antibioticum Violamycin BI.** Z. Allg. Mikrobiol., **17,** 569–573 (1977) (KOM).

The photodynamic action of violamycin BI complex (V) is small compared with the photodynamic active dye methylene blue (MB), using *P. mirabilis* as test organism. Furthermore, with increasing illumination time the photodynamic action of V decreases. Therefore, the kinetics of inactivation of bacteria obtained with V is different from that obtained with MB. The repair of photodynamically produced DNA damages (MB as sensitizer) is not inhibited, if concentrations are used having only small or no inhibitory effect on photodynamically treated bacteria.

ZEITSCHRIFT FÜR ALLGEMEINE MIKRO-BIOLOGIE

MORPHOLOGIE, PHYSIOLOGIE, GENETIK UND ÖKOLOGIE DER MIKROORGANISMEN

HEFT 7 · 1977 · BAND 17

AKADEMIE-VERLAG · BERLIN

Die Zeitschrift für Allgemeine Mikrobiologie soll dazu beitragen, Forschung und internationale Zusammenarbeit auf dem Gebiet der Mikrobiologie zu fördern. Es werden Manuskripte aus allen Gebieten der allgemeinen Mikrobiologie veröffentlicht. Arbeiten über Themen aus der medizinischen, landwirtschaftlichen, technischen Mikrobiologie und aus der Taxonomie der Mikroorganismen werden ebenfalls aufgenommen, wenn sie Fragen von allgemeinem Interesse behandeln.

Zur Veröffentlichung werden angenommen:

Originalmanuskripte, die in anderen Zeitschriften noch nicht veröffentlicht worden sind und in gleicher Form auch nicht in anderen Zeitschriften erscheinen werden. Der Umfang soll höchstens 1½ Druckbogen (24 Druckseiten) betragen. Bei umfangreicheren Manuskripten müssen besondere Vereinbarungen mit der Schriftleitung und dem Verlag getroffen werden.
Kurze Originalmitteilungen über wesentliche, neue Forschungsergebnisse. Umfang im allgemeinen höchstens 3 Druckseiten. Kurze Originalmitteilungen werden beschleunigt veröffentlicht.
Kritische Sammelberichte und Buchbesprechungen nach Vereinbarung mit der Schriftleitung

Bezugsmöglichkeiten der Zeitschrift für Allgemeine Mikrobiologie:

Bestellungen sind zu richten
- **in der DDR** an eine Buchhandlung oder an den Akademie-Verlag, DDR-108 Berlin, Leipziger Straße 3–4
- **im sozialistischen Ausland** an eine Buchhandlung für fremdsprachige Literatur oder an den zuständigen Postzeitungsvertrieb
- **in der BRD und Westberlin** an eine Buchhandlung oder an die Auslieferungsstelle KUNST UND WISSEN, Erich Bieber, 7 Stuttgart 1, Wilhelmstraße 4–6
- **in Österreich** an den Globus-Buchvertrieb, 2101 Wien, Hochstädtplatz 3
- **im übrigen Ausland** an den Internationalen Buch- und Zeitschriftenhandel; den Buchexport, Volkseigener Außenhandelsbetrieb der Deutschen Demokratischen Republik, DDR-701 Leipzig, Postfach 160, oder an den Akademie-Verlag, DDR-108 Berlin, Leipziger Straße 3–4.

Zeitschrift für Allgemeine Mikrobiologie

Herausgeber: Im Auftrag des Verlages von einem internationalen Wissenschaftlerkollektiv herausgegeben.
Verlag: Akademie-Verlag, DDR-108 Berlin, Leipziger Straße 3–4; Fernruf 2236221 oder 2236229; Telex-Nr. 114420; Postscheckkonto: Berlin 35021, Bank: Staatsbank der DDR, Berlin, Kto.-Nr.: 6836-26-20712.
Chefredaktion: Prof. Dr. Udo Taubeneck,
Prof. Dr. Wilhelm Schwartz.
Anschrift der Redaktion: Zentralinstitut für Mikrobiologie und experimentelle Therapie der Akademie der Wissenschaften, DDR-69 Jena, Beuthenbergstr. 11; Fernruf: Jena 885614; Telex-Nr. 058621.
Veröffentlicht unter der Lizenznummer 1306 des Presseamtes beim Vorsitzenden des Ministerrates der Deutschen Demokratischen Republik.
Gesamtherstellung: VEB Druckerei „Thomas Müntzer“, DDR-582 Bad Langensalza.
Erscheinungsweise: Die Zeitschrift für Allgemeine Mikrobiologie erscheint jährlich in einem Band mit 8 Heften. Bezugspreis je Band 160,– M zuzüglich Versandspesen. Preis je Heft 20,– M.

Erscheinungstermin: Oktober 1977

Bestellnummer dieses Heftes 1070/17/7

AN (EDV) 75218

Zeitschrift für Allgemeine Mikrobiologie	17	7	1977	501–506

(Department of Microbiology, University of Hawaii, Honolulu, HI 96822, USA)

Dissimilatory nitrate reduction in *Clostridium tertium*

M. HASAN[1]) and J. B. HALL

(Eingegangen am 8. 9. 1976)

Fermentation balance studies were carried out on *Clostridium tertium* grown with and without nitrate in the medium. Nitrate reduction increased the efficiency of energy produced from glucose by permitting the utilization of additional sites of substrate level phosphorylation. The effect was even more dramatic in *C. tertium* than in *C. perfringens*, with increased cell yields of about 30% being observed in the former compared with 20% in the latter.

Unlike *C. perfringens*, *C. tertium* responded to the presence of nitrate in the medium with an increased growth rate. A slight increase in the Y_{ATP} of these cultures was also observed, and quantitatively, this appeared to be consistent with the prediction of STOUTHAMMER and BETTENHAUSSEN that Y_{ATP} will vary with the growth rate.

Thus, *C. tertium*, like *C. perfringens*, was able to use nitrate as an electron acceptor in conjunction with its energy metabolism, suggesting that this trait may be widespread among the nitrate-reducing anaerobes.

A previous report (HASAN and HALL 1975) presented evidence that *Clostridium perfringens* was able to increase the amount of ATP produced per mole of glucose consumed when nitrate was available to serve as an electron acceptor, and thus could be considered to engage in a rudimentary form of nitrate respiration. This anaerobe is unusually aerotolerant for a member of its genus, and may not be typical in its energy metabolism. Since a significant role for this primitive mode of nitrate respiration in the initiation of the evolution of respiratory and photosynthetic systems has been postulated (HALL, 1971, 1973), it seemed important to examine other nitrate-reducing members of this genus to determine whether or not nitrate reduction could play a role in their energy metabolism also. *Clostridium tertium*, which is assigned to a different subdivision of the genus than *C. perfringens* (SMITH and HOBBS 1974), was chosen for this study.

Materials and methods

Cultures and growth conditions: *Clostridium tertium* No. 2771 from the collection of the VPI Anaerobe Laboratory (Virginia Polytechnic Institute and State University, Blacksburg, Virginia, U.S.A.) was used in these experiments. The bacteria were grown on a partially defined medium based on that of FUCHS and BONDE (1957) in 13 × 100 mm screw capped tubes each containing 5 ml of culture. Anaerobic conditions were maintained by a layer of mineral oil on top of the medium and by the presence of ascorbate in it. An inoculum was prepared by harvesting an actively growing culture of *C. tertium* by centrifugation at 9000 g for 15 min, aseptically washing the bacteria twice with 0.1 M phosphate buffer, pH 7.0, and resuspending them in one-half the original volume. These cells were then used to inoculate the culture tubes in each experiment.

Growth was measured by following the absorbancy at 600 nm and converting the readings to dry weights by reference to a standard curve.

Assay of fermentation products: All fermentation products were determined by the methods described in a previous paper (HASAN and HALL 1975).

[1]) Present address: Dept. of Biological Chemistry, University of Maryland, School of Medicine, Baltimore, Maryland 21201, USA

Results

Growth rate and growth yields

When the growth rates of *C. tertium* on the medium, with and without added nitrate, were compared, a significantly greater rate was observed in the presence of nitrate (Fig. 1).

Fermentation balance experiments were carried out in media containing 400 to 500 mg/l of glucose, a concentration which was limiting for growth as shown in Fig. 2. As this figure also indicates, the yield of cells in the presence of 1.6 µmoles/ml of

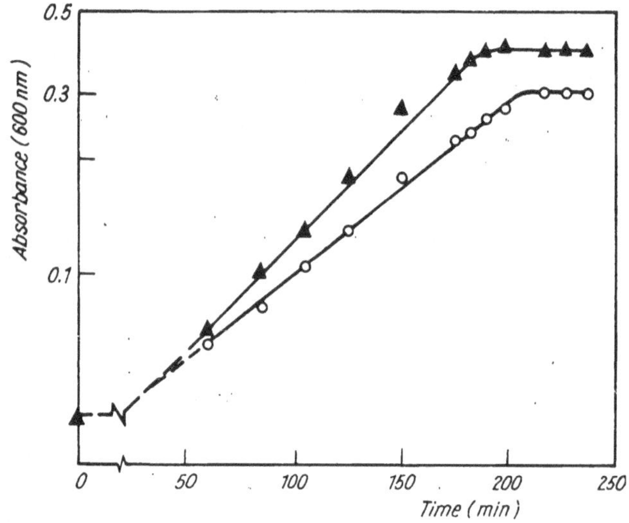

Fig. 1. Growth of *C. tertium* on partially defined medium in the presence of nitrate (▲) and in the absence of nitrate (○). Growth was measured as A_{600}

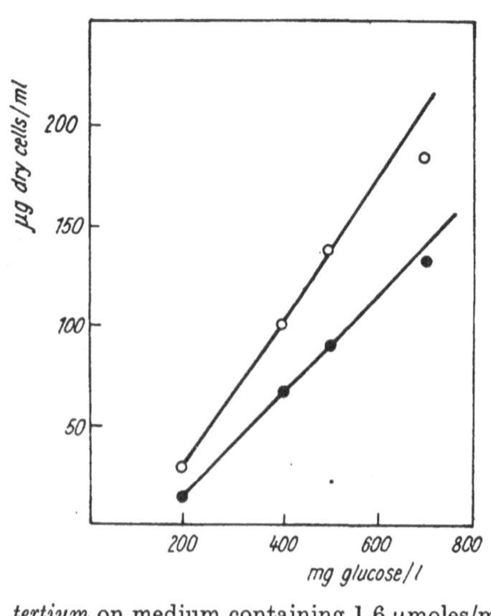

Fig. 2. Growth yields of *C. tertium* on medium containing 1.6 µmoles/ml of nitrate (○) and on medium lacking nitrate (●)

nitrate was quite substantially greater than in its absence at all glucose concentrations tested. These bacteria achieve maximum yield on appreciably lower concentrations of glucose than does *C. perfringens* (Hasan and Hall 1975) but, as in the case of the latter, their growth was strictly dependent on the presence of this sugar and ceases when its concentration becomes too low despite the wealth of other nutrients in the medium.

Fate of nitrate in the cultures

Nitrite was an intermediate in nitrate reduction and accumulated in the culture media of many clostridia growing in the presence of nitrate (Reed 1942). Nitrite is known to be toxic to most bacteria and in *C. perfringens* the ability of nitrate to support enhanced growth yields was limited to a rather narrow range of concentrations, probably due to the accumulation of nitrite (Hasan and Hall 1975). *C. tertium* accumulated nitrite during growth on nitrate, but within the range that was examined growth did not seem to be inhibited by this ion (Table 1). Most of the nitrite was further reduced to ammonium by this organism. As shown in Table 2, none of this ammonium seemed to be assimilated by the bacteria, nor was additional ammonium produced from the organic constituents of the medium. *C. perfringens* has an absolute requirement for ammonium salts (Fuchs and Bonde 1957) and reducing the amount present to that which was found as a contaminant in the amino acid mixture resulted in a significant reduction in the growth rate. *C. tertium* showed no such sensitivity to the availability of free amonium, however, and presumably was able to satisfy its entire nitrogen requirements from the organic compounds in the medium.

Table 1
Growth of *C. tertium* on media containing different concentrations of nitrate

Initial nitrate in medium[1])	Nitrite at end of experiment[1])	Growth yield[2])
0	0	91
1.6	0.13	130
3.2	1.63	135
4.8	1.84	130
6.5	1.84	138

[1]) Expressed in μmoles/ml
[2]) Expressed in μg dry weight/ml

Table 2
Inorganic nitrogen balances of cultures grown in media with no added ammonium salts

Compound	Amounts measured[1])		
	Experiment 1	Experiment 2	Control
Nitrate added	1.56	1.56	0.00
Nitrate remaining	0.28	0.23	—
Nitrate used	1.28	1.33	—
Nitrite remaining	0.034	0.04	0.00
Initial ammonium in medium[2])	2.9	2.9	2.9
Total ammonium found after bacterial growth	4.2	4.2	2.9
Ammonium produced from nitrate	1.3	1.4	0.00
Nitrate-nitrogen accounted for	1.6	1.67	—
Percent recovered	102	109	

[1]) Expressed as μmoles/ml
[2]) This ammonium was probably a contaminant of the amino acid mixture

Fermentation balances

Fermentation balances for *C. tertium* grown with and without nitrate are given in Table 3. The shift in metabolism in the presence of nitrate was quite marked. The amount of acetate produced more than doubled while butyrate production was only one-tenth that of the nitrate-free cultures, and ethanol, which was a major product in the latter was completely absent when nitrate was available. The amount of hydrogen produced when nitrate was present also declined to about a third of the value in the absence of nitrate. Again, as observed previously, there was a shift

Table 3
Fermentation of glucose in the presence and absence of nitrate by *C. tertium*[1])

	With nitrate		Without nitrate	
Determination	1	2	1	2
Glucose used (µmoles/ml)	2.68	2.6	2.68	2.6
Products (moles/moles glucose)				
Acetate	1.63	1.62	0.64	0.67
Butyrate	0.02	0.07	0.21	0.25
Lactate	0.17	0.22	0.20	0.23
Ethanol	0.00	0.00	0.61	0.46
Carbon dioxide	—	1.56	—	1.73
Hydrogen	—	0.58	—	1.51
Nitrite	0.00	0.00	—	—
Nitrate used	0.58	0.58	—	—
Dry weight of cells (µg/ml)	123.5	—	91.5	—
Grams dry weight/mole glucose	46.0	—	34.1	—
Carbon recovered (percent)	(90)	95	(94)	95
O/R index	(1.05)	1.04	(1.07)	1.15
C-1/C-2 ratio	(0.93)	0.89	(1.03)	1.06

[1]) Each value is the average of two determinations, except for the gas values, each of which required a pair of flasks for the assay. Values in parentheses were calculated by supplying the missing analytical values from the average values (mole/mole glucose) found in the relevant parallel experiments. The ratio of the relatively oxidized organic compounds to reduced products (O/R index) was calculated by assuming that all of the nitrate consumed was converted to nitrite (measured) and ammonia (the remainder).

Table 4
Increases in ATP production and cell yields per mole of glucose consumed in cultures of *C. tertium* and *C. perfringens* grown in the presence of nitrate compared with those grown in its absence

Condition	*C. tertium*	*C. perfringens*
Average ATP produced (mole/mole glucose)		
With nitrate	3.58	3.08
Without nitrate	2.75	2.58
Difference	0.80	0.50
Percent increase with nitrate	29	19
Average cell yield (g/mole glucose)		
With nitrate	46.0	45.0
Without nitrate	34.1	38.3
Difference	11.9	6.7
Percent increase with nitrate	35	17
Average Y_{ATP} (g dry wt bacteria produced per mole ATP generated)		
With nitrate	13.1	14.6
Without nitrate	12.3	14.8

from the production of more reduced products to acetate which permitted additional ATP synthesis via acetyl-phosphate. The resulting increase in ATP yield, calculated on the assumption that 2 moles ATP were produced per mole acetate, 3 per mole butyrate, and 1 each per mole ethanol and lactate is indicated in Table 4.

Discussion

Although an increase in yield in the presence of nitrate was observed in *C. perfringens* over the yield in the absence of nitrate, no acceleration in the rate of growth of the cultures was noted (HASAN and HALL 1975). Apparently, the rate of energy production in these cultures was not the limiting factor for the rate of growth, or the presence of nitrate, though permitting more efficient utilization of glucose, did not increase the rate of energy production from it. *C. tertium*, however, did grow more rapidly in the presence of nitrate as well as yielding a greater dry weight of cells per mole of glucose utilized. The molar growth yield per mole ATP, Y_{ATP}, was also slightly higher in the nitrate cultures. STOUTHAMMER and BETTENHAUSSEN (1973) have argued that Y_{ATP} is not a constant but is influenced by the composition of the cells, the specific growth rate, and the maintenance coefficient. They present a graph (Fig. 3 in the reference cited) relating Y_{ATP} and specific growth rates at various values of the maintenance coefficient, *m*. If the average Y_{ATP} for *C. tertium* cultures grown with and without nitrate, 13.1 and 12.3 (from Table 4), are plotted against the specific growth rates of 0.83 and 0.64 per hour, respectively (from Fig. 1), both points fall on the curve representing a maintenance coefficient of about 28 gmoles ATP/g dry weight per h. Thus, our data are consistent with the hypothesis of STOUTHAMMER and BETTENHAUSEN, which suggests that the slight apparent increase in Y_{ATP} observed in the presence of nitrate can be attributed to the increased growth rate.

A comparison between the increased cell yields and ATP produced per mole of glucose in the presence of nitrate by *C. tertium* with the same figures published earlier (HASAN and HALL 1975) for *C. perfringens* (Table 4) revealed that *C. tertium* obtained even more benefit by nitrate reduction than *C. perfringens*. This may have been due to the differing growth conditions, however, particularly to the greater amount of nitrate relative to glucose available to the former organism. Glucose, nitrate, and nitrite concentrations varied continually during growth of these cultures and so the estimates of the advantage of nitrate reduction to the bacteria obtained from these batch cultures probably understate the benefit obtainable under optimum conditions with each constituent maintained at the proper constant level.

ISHIMOTO *et al.* (1974) recently published studies similar to ours on the effect of nitrate on the fermentation products of *C. perfringens*. They observed a similar effect on the distribution of fermentation products, but found no increase in growth yield in the presence of nitrate. This was not inconsistent with our results, however, as we had reported that the increased growth yields could be observed only within a relatively narrow range of nitrate concentrations, with an optimum at about 3.2 μmoles/ml, and that somewhere between 5 and 10 μmoles/ml the effect vanished, probably due to the adverse effects of the nitrite that accumulated in these cultures. As they used nitrate concentrations of 20 μmoles/ml, it is not surprising that no enhancement of yields were observed.

Finally, it should be emphasized that the nitrate reduction observed in cultures of *C. tertium* appears to be dissimilatory reduction and not assimilatory in character. This is supported by the following observations: (1) The bacteria did not appear to require an inorganic nitrogen source in the medium but satisfied their requirements for nitrogen entirely from the amino acids and other organic compounds that they did require. (2) The ammonium that was produced by nitrate reduction was not assimilated by the cells (Table 2). (3) When ammonium salts are supplied in the medium in concentrations more than adequate to satisfy all cell requirements, no repression of the nitrate-reducing system was observed. These observations reinforce our impression that the nitrate-reducing system is retained by these bacteria for its value in energy metabolism. The finding of a second species of *Clostridium* which reduced nitrate as a way of increasing the efficiency with which it could produce ATP from glucose suggests that this primitive form of respiration may be fairly widespread among nitrate-reducing anaerobes.

Acknowledgements

We thank Dr. T. D. WILKINS and his colleagues at the Virginia Polytechnic Institute Anaerobe Laboratory for supplying us with bacterial cultures and much valuable information about anaerobic techniques. This work was supported by funds of the Department of Microbiology of the University of Hawaii.

References

FUCHS, A. R. and BONDE, G. J., 1957. The nutritional requirements of *Clostridium perfringens*. J. gen. Microbiol., **16**, 317–329.
HALL, J. B., 1971. Evolution of the prokaryotes. J. theor. Biol., **30**, 429–454.
HALL, J. B., 1973. The occurrence of nitrate on the early earth and its role in the evolution of the prokaryotes. Space Life Sciences, **4**, 204–213.
HASAN, S. M. and HALL, J. B., 1975. The physiological function of nitrate reduction in *Clostridium perfringens*. J. gen. Microbiol., **87**, 120–128.
ISHIMOTO, M., UMEYAMA, M. and CHIBA, S., 1974. Alteration of fermentation products from butyrate to acetate by nitrate reduction in *Clostridium perfringens*. Z. Allg. Mikrobiol., **14**, 115–121.
REED, R. W., 1942. Nitrate, nitrite, and indole reactions of gas gangrene anaerobes. J. Bacteriol., **44**, 425–431.
SMITH, L. D. S. and HOBBS, G., 1974. *Clostridium*. In: BERGEY's Manual of Determinative Bacteriology (Eighth Ed. pp. 551–572; R. E. BUCHANAN and N. E. GIBBONS, Editors). Williams and Wilkins Co. Baltimore, Maryland, U.S.A.
STOUTHAMER, A. H. and BETTENHAUSSEN, C., 1973. Utilization of energy for growth and maintenance in continuous and batch cultures of microorganisms — a reevaluation of the method for the determination of ATP production by measuring molar growth yields. Biochim. biophysica Acta, **301**, 53–70.

Mailing address: Dr. J. B. HALL
Department of Microbiology, University of Hawaii,
2538 The Mall Honolulu, HI 96822, USA

Zeitschrift für Allgemeine Mikrobiologie	17	7	1977	507–512

(Department of Applied Chemistry, Faculty of Science, Science University of Tokyo, Kagurasaka, Shinjuku-ku, Tokyo, Japan)

Introduction of a 16α-hydroxyl function into estrone by *Streptomyces roseochromogenes*

M. IIDA and H. IIZUKA

(Eingegangen am 3. 8. 1976)

Streptomyces roseochromogenes (NRRL B-1233) converted estrone predominantly into its 16α-hydroxyl derivative. Chemical and spectroscopic (UV, IR, NMR, MS) methods were used in establishing the structure and stereochemistry of the oxidation product. The product was assigned as 16α-hydroxyestrone (yield, 17%). No other oxidation product was detected in this experiment. An interrelationship between cell growth and 16α-hydroxy-estrone formation was observed. Also, 16α-hydroxylation of estrone was observed in resting cells.

16α-Hydroxylase showed good activity at 3.7×10^{-4}M of estrone concentration and was completely inhibited by 1.1×10^{-3}M. These results indicate the presence of a constitutive 16α-hydroxylase in the organism investigated.

Many investigators have reported the hydroxylation of steroids by microorganisms (CHARNEY and HERZOG 1967). The introduction of a 16α-hydroxyl function in a steroid nucleus is one of the important reactions in mammalian biosynthesis of estriol. Since 1973, we have studied the microbiological 16α-hydroxylation of the possible intermediates in estriol biosynthesis. In previous papers, we have reported the microbiological formation of 16α-hydroxy-dehydroepiandrosterone (16α-OH-DHEA) as the main precursor of estriol (IIDA *et al.* 1975a) and ^{3}H-labelled 16α-OH-DHEA having high specific activity (specific activity, 13.9 Ci/mM) (IIDA *et al.* 1975b). Little is known of microbiological 16α-hydroxylation of estrogens (KITA *et al.* 1961).

The present paper deals with the microbiological 16α-hydroxylation of estrone by *Streptomyces roseochromogenes*.

Materials and methods

Cultivation: *Streptomyces roseochromogenes* NRRL B-1233 was the most appropriate microorganism for our purpose. *S. roseochromogenes* was grown in 500 ml conical flasks containing 100 ml of the medium (glucose 3%, soybean oil 0.22%, soybean meal 2%, $CaCO_3$ 0.25%, pH 7.2) on a rotary shaker (220 rpm) at 27 °C. At the early exponential growth phase, 10 mg of estrone dissolved in 0.5 ml of N,N-dimethylformamide was added and the incubation was continued for additional 48 hrs.

Substrate used: Estrone was purchased from IKAPHARM Ltd., Ramat-Gam. P.O.B. 31, Israel.

Isolation of oxidation product: The reaction mixture was extracted three times with successive equivalents of methylene chloride. The combined extracts were evaporated *in vacuo*. The residue was dissolved in ethanol and subjected to chromatography on a silica gel column with 1% ethanol-methylene chloride as the eluent. The fraction containing the oxidation product was chromatographed on preparative silica gel thin-layer plates in chloroform-methanol (97:3, v/v). The absorbent corresponding to the spot (Rf 0.2) was scraped off and eluted with three 10 ml portions of the solvent (ethyl acetate:*n*-hexane = 2:1 v/v). The product obtained was twice recrystallized from ethanol.

Analysis of oxidation product: Gas liquid chromatographic (GLC) analysis was performed with a SHIMADZU GC-4B PTF chromatographic unit, using nitrogen (1 kg/cm²) as the carrier gas and

1.5% OV-1 on 1.5 m × 4 mm column. The column temperature was 210 °C and the detector temperature 240 °C. The ultraviolet (UV) absorbtion spectrum was measured in ethanol using a Shimadzu model SV-50A. Infrared (IR) absorption spectrum was obtained with a Nippon Bunko model DS-301 spectrophotometer in KBr disk. Mass (MS) spectral-analysis was carried out on a Hitachi model RMU-6L instrument (Chamber voltage 70 eV). Nuclear magnetic resonance (NMR) determination was made with a Jeol model JNM-4H100 instrument. Sample in deuterioacetone was analyzed by using tetramethylsilane as an internal standard. The product was determined by quantitative gas chromatography.

The experiments were carried out in triplicate and the results of this work are the arithmetic mean of these replicates.

Results and discussion

Suitability of the culture medium for the monohydroxyl derivative

The reaction mixture was presumed the presence of monohydroxyl derivative using TLC and GLC. The other oxidation products were not detected on any media. The formation of the monohydroxyl derivative on various media is summarized in Fig. 1. The formation of the monohydroxyl derivative was observed on medium I through III. The yields were 4.6%, 17.8%, and 8%, respectively. The yield of monohydroxyl derivative on medium II was greater than those of any other media. On the other hand, all media favoured the formation of monohydroxylase.

Medium II was used throughout the following experiments.

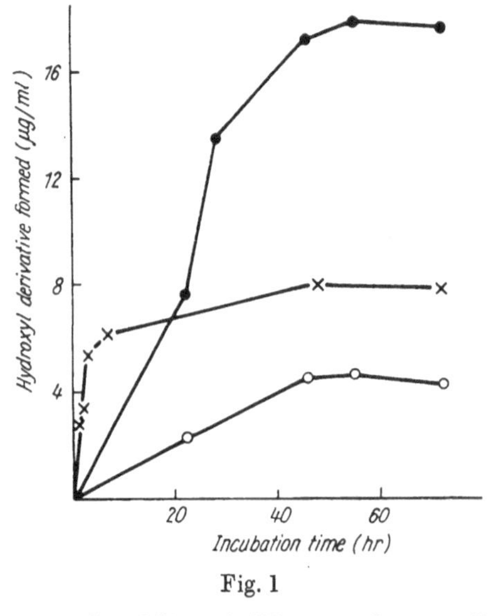

Fig. 1

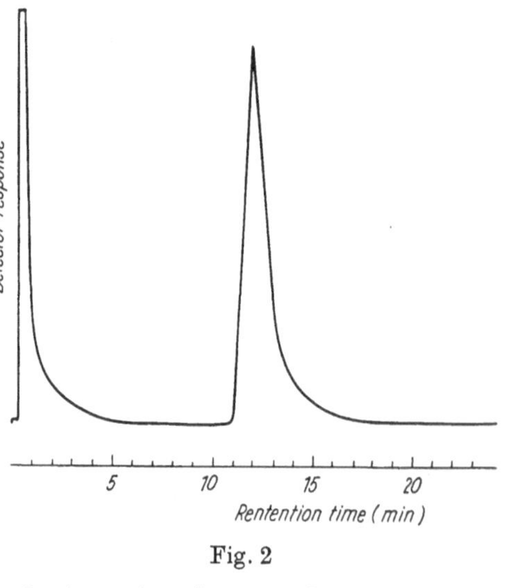

Fig. 2

Fig. 1. Suitability of different culture media for the formation of monohydroxyl derivative. Composition of media (g/l): ○ Medium I: glucose 10, polypeptone 2, meat extract 1, yeast extract 1, glycerine 0.5, pH 7.2. ● Medium II: glucose 30, soybean meal 20, soybean oil 2.2, $CaCO_3$ 2.5, pH 7.2. × Medium III: glucose 10, soybean meal 2.2, corn steep liquor 3, yeast extract 2.5, $NH_4H_2PO_4$ 3, $CaCO_3$ 2.5, pH 7.2

Fig. 2. Gas chromatogram of oxidation product obtained from estrone. GLC was performed with a Shimadzu GC-4B PTF on 1.5 m × 4 mm column. The column temperature was 210 °C and the detector temperature 250 °C

16α-Hydroxy-estrone formed from estrone

The chromatograms of the oxidation are summarized in Table 1. As shown clearly, the oxidation product exhibited the same TLC and GLC behaviour as the authentic 16α-hydroxyestrone (16α-OH-estrone). Mp. 204—206 °C (uncorrect). The calculated values for $C_{18}H_{22}O_3$ were 75.49% C, 7.74% H, the experimental values being 75.55% C, 7.61% H. MS of the product showed a peak of parental molecular ion M^+ at m/e 286. As shown in Fig. 2, the retention time of the product was 12.0 min. UV and IR spectra of the product are shown in Figs. 3. and 4, respectively. UV λmax (EtOH)

Table 1
Chromatographic properties of oxidation product obtained from estrone

	Rf-value on TLC[1]) 1	2	Retention time[2]) on OV-1
Estrone	0.39	0.46	9.0
16 α-OH-estrone	0.20	0.27	12.7
Oxidation product	0.20	0.27	9.0
	0.39	0.46	12.7

[1]) Solvent system: 1, chloroform: methanol (97:3, v/v);
2, cyclohexane: AcOEt (1:1, v/v)

[2]) GLC analysis was performed with a SHIMADZU model GC-4B PTF chromatographic unit, using on 1.5 m × 4 mm column. The column temperature was 210 °C and the detector temperature 240 °.

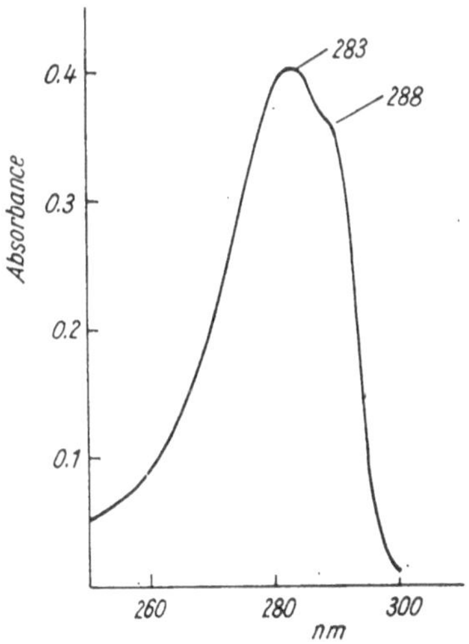

Fig. 3. UV-spectrum of oxidation product obtained from estrone (in EtOH). 283 (ε 1835), 288 (ε 1659)

nm: 283 (ε 1835), 288 shoulder (ε 1659); IR νmax (0.6% KBr) cm^{-1}: 3360 (OH), 1730 (17—CO), 1615, 1658, 1500 1455 (aromatic C = C). NMR of the product is shown in Fig. 5. NMR 100 MHz, 30% solution in $(CD_3)_2CO$ ppm: 0.97 (C_{18}3H, s), 1.2—2.5 (11H), 2.7—2.9 ($C_{5,6}$ 2H, m), 4.17 (C_{16}OH, 1H, d, J = 4 Hz), 4.34 (C_{16} H, 1H, m), 6.54 (C_4 1H, broad, s), 6.58 (C_2, 1H, d, d, J = 8.2 Hz), 7.08 (C_1 1H, d, J = 8 Hz), 7.89 (C_3 OH, 1H, s). On addition of D_2O (Fig. 6), the doublet of a hydroxyl group (1H, 4.17 ppm) at C_{16} and the singlet of a hydroxyl group (1 H, 7.89 ppm) at C_3 disappeared.

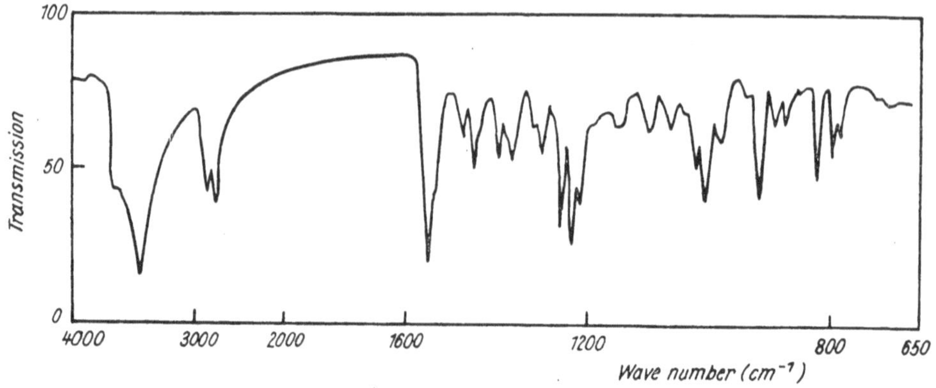

Fig. 4. IR-spectrum of oxidation product obtained from estrone (in 0.6% KBr)

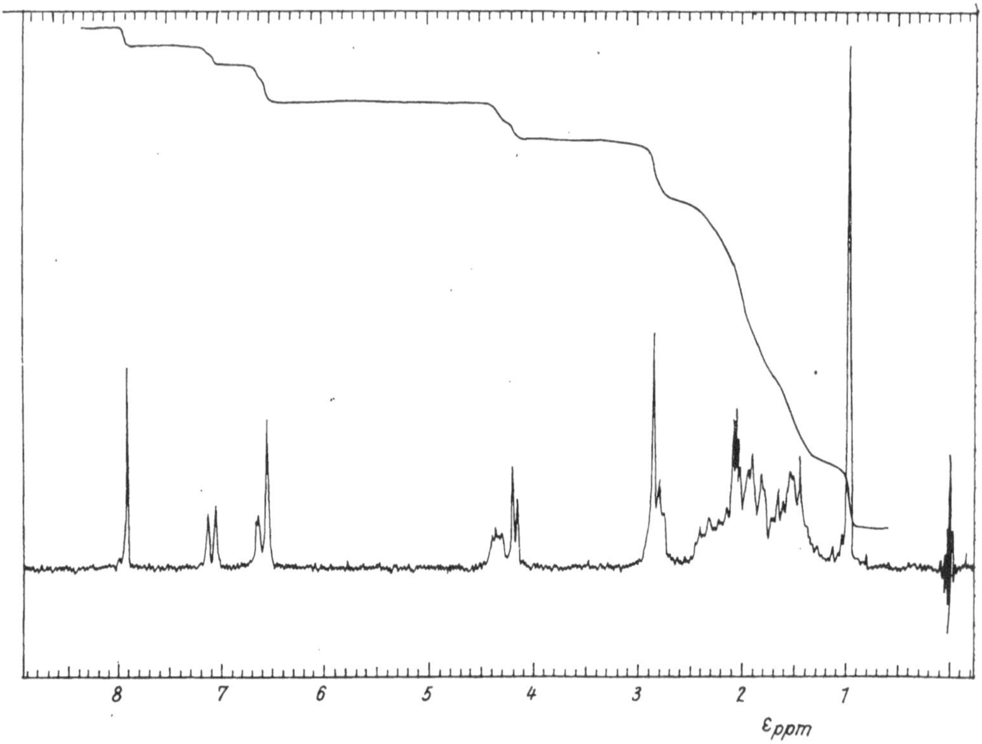

Fig. 5. NMR-spectrum of oxidation product obtained from estrone

On the basis of the above mentioned results the product obtained was assigned as the structure of 16α-OH-estrone. 16α-OH-estrone was obtained in yield of above 17 mg.

Relationship between the cell growth and the formation of 16α-OH-estrone

The formation of 16α-OH-estrone as a function of time was examined. The result is shown in Fig. 7. The formation of 16α-OH-estrone seems to have a close relation

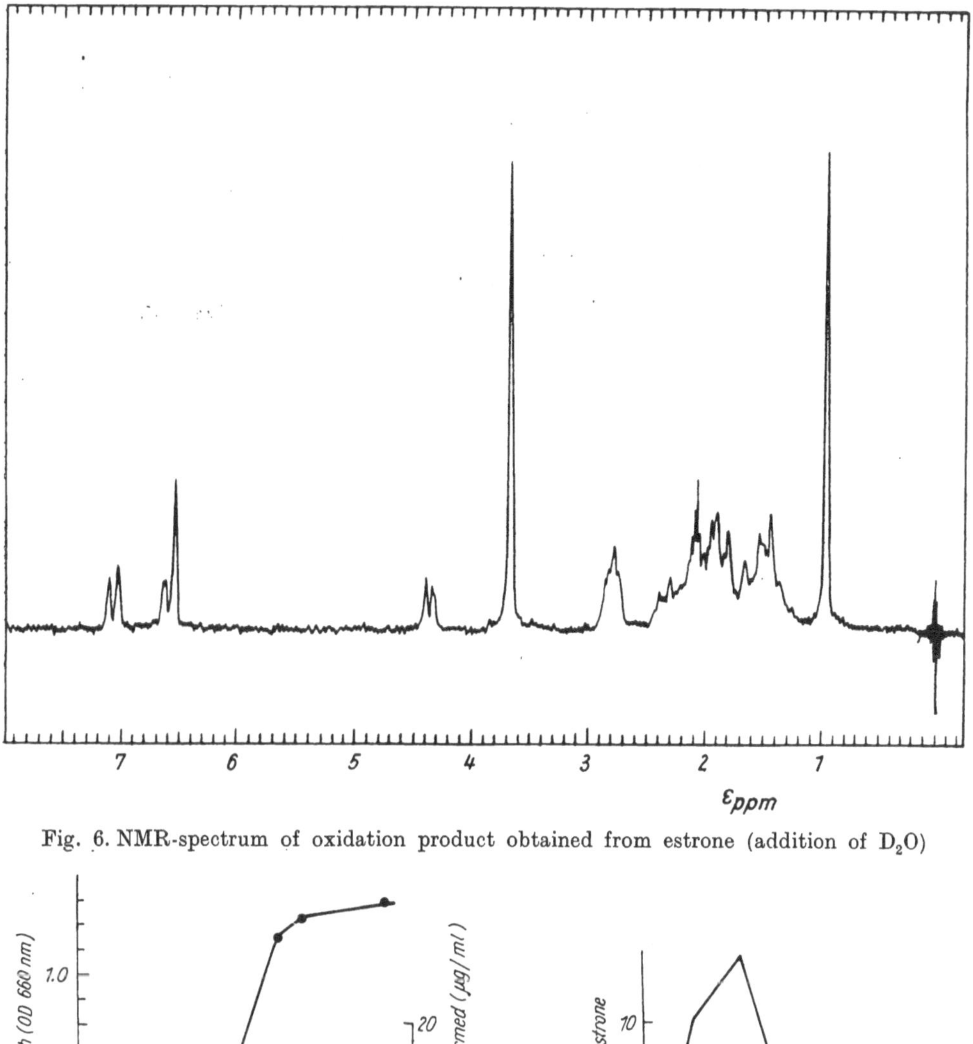

Fig. 6. NMR-spectrum of oxidation product obtained from estrone (addition of D_2O)

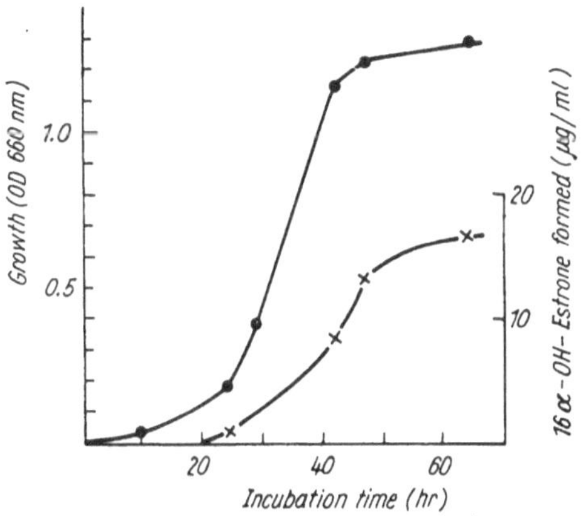

Fig. 7. Relationship between cell growth (●) and formation of 16α-OH-estrone (×)

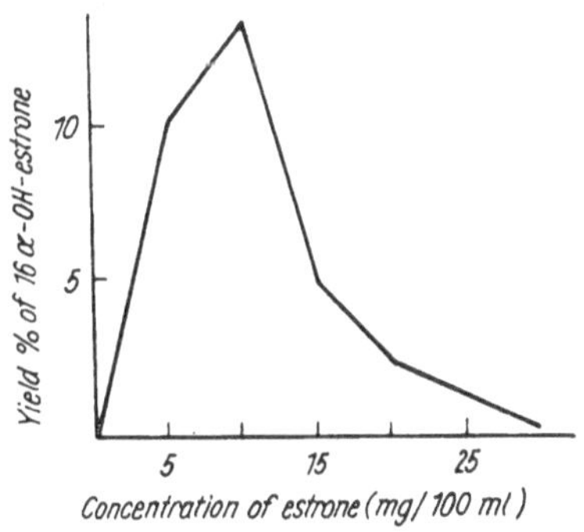

Fig. 8. 16α-hydroxylation activity as a function of estrone concentration

to cell growth and reached the maximum (17 μg/ml) after approximately 48 hrs and then gradually decreased with the passage of cultivation. 16α-hydroxylation activity as a function of estrone concentration is shown in Fig. 8. 16α-hydroxylation showed good activity at 3.7×10^{-4} M of estrone and completely was inhibited by $1.1. \times 10^{-3}$ M.

The 16α-hydroxylation of estrone by resting cells was also investigated. The cells were harvested at the late exponential growth phase by centrifugation at 25,500 × g for 10 min, washed twice with cold 0.2 M Tris-HCl buffer (pH 7.2) containing 0.001% $FeSO_4$ and resuspended in the same buffer at the concentration of approximately 1.85 g wet wt. per 100 ml. The suspension was shaken at 27 °C for 26 hrs in the presence of 5 mg estrone. The product exhibited a chromatographic behaviour identical to those of 16α-OH-estrone. No other oxidation product was detected. The formation of 16α-OH-estrone as a function of time is shown in Fig. 9. 11% estrone were trans formed to 16α-OH-estrone after approximately 21 hrs of incubation. From these findings, the strain investigated is presumed to possess a constitutive 16α-hydroxylase.

This paper was presented at the Fifth International Fermentation Symposium in Berlin (West), on June 30, 1976.

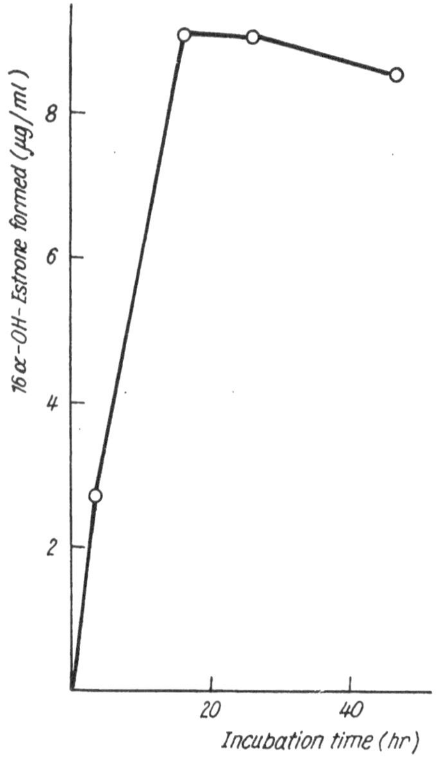

Fig. 9. Formation of 16α-OH-estrone by resting cells

References

CHARNEY, W. and HERZOG, H. L., 1967. Microbial Transformation of Steroids — A Handbook. Academic Press New York and London.

IIDA, M., MATSUHASHI, K. and NAKAYAMA, T., 1975a. Microbial oxidation of dehydroepiandrosterone and related compounds. Z. Allg. Mikrobiol., **15**, 181—187.

IIDA, M., MATSUHASHI, K. and NAKAYAMA, T., 1975b. Biosynthesis of [7-^{3}H] 16α-hydroxy-dehydroepiandrosterone having high specific activity. Z. Allg. Mikrobiol., **15**, 189—193.

KITA, D. A., SARDINAS, J. L. and SHULL, G. M., 1961. Introduction of the 16α-hydroxyl group into estrogens by streptomycetes. Nature, **190**, 627—628.

Mailing address: Dr. M. IIDA
Department of Applied Biological Science,
Faculty of Science and Technology,
Science University of Tokyo,
Noda, Chiba, Japan

Zeitschrift für Allgemeine Mikrobiologie	17	7	1977	513–519

(Department of Botany, Banaras Hindu University, Varanasi, India)

Some characteristics of two morphological mutants of *Nostoc linckia* induced by nitrosoguanidine

J. K. LADHA and H. D. KUMAR

(Eingegangen am 1. 9. 1976)

The blue-green alga *Nostoc linckia* was treated with nitrosoguanidine and two classes of morphological mutant clones were isolated. One class shows certain abnormal phenotypic features of vegetative cells, spores, and heterocysts. It has increased heterocyst frequency and impaired growth rate. The other class exhibits an altered heterocyst spacing pattern. Both classes of mutants have reduced nitrogenase activity.

The production of filamentous mutants following treatment with mutagenic agents has been reported in bacteria (GREENBERG *et al.* 1961) and in blue-green algae (KUMAR 1963, VAN BAALEN 1973). INGRAM and VAN BAALEN (1970) reported a stable filamentous mutant in a coccoid blue-green alga *Agmenellum quadruplicatum* strain BG 1 after treatment with nitrosoguanidine. SINHA and KUMAR (1973) obtained a non-nitrogen fixing mutant of *Anabaena doliolum* and also described the existence of several abnormal and bizarre forms in this mutant strain. However, no quantitative study of the incidence of such morphological aberrations was made.

We have isolated two classes of morphological mutant clones of *Nostoc linckia* following its treatment with nitrosoguanidine. One class of mutant shows certain abnormal phenotypic features of vegetative cells, spores, and heterocysts. It has increased heterocyst frequency and impaired growth rate. The other class of mutant exhibits an altered heterocyst spacing pattern.

Materials and methods

Organism: *Nostoc linckia* was collected locally from rice-paddies of Banaras Hindu University. Natural populations of this species were found to be mostly unialgal. It was homogenized by shaking with sterile glass beads in 10 ml of sterile nitrogen free medium. A very thin suspension was streaked on agar plates and single-celled or single-spored colonies were picked off as described in our previous paper (KUMAR and LADHA 1976).

Nostoc linckia (ROTH.) BORN. *et* FLAH. is a nitrogen-fixing filamentous and sporulating blue-green alga which readily forms discrete clones on agar medium.

Culture medium: The alga was grown under unialgal and bacteria-free conditions on ALLEN and ARNON's nitrogen-free medium (pH 7.5) (for composition of the medium and other growth conditions, see LADHA and KUMAR 1975). ALLEN and ARNON's nitrogen-free medium supplemented with 1 mM of ammonium chloride was also used in some experiments.

Isolation of mutants: Exponential-phase cultures of wild-type algae grown in batch culture in liquid nitrogen-free medium, were treated with 1-methyl-3-nitro-nitrosoguanidine (NTG) (100 μg/ml) for 30 min in the dark at room temperature. The algal suspension was continuously agitated during treatment. This treatment permitted about 30% survival. Cells were washed with sterile medium by centrifugation and inoculated into nitrogen-free liquid medium. They were then incubated under optimal growth conditions for 4–5 days. Samples were then removed and inocula were streaked on agar plates containing nitrogen-free medium with 1% DIFCO agar. After a 7–9 day incubation period, presumed variant colonies were picked off and inoculated separately into 5 ml of nitrogen-free medium. The variant clones were repeatedly subcultured serially in nitrogen-free medium for several months. Those which proved stable were once again streaked on agar plates, incubated, and single colonies picked up for further characterization.

Estimation of nitrogenase activity: Nitrogenase activity was assayed by the acetylene reduction techniques at the Department of Botany and Biochemistry, Westfield College, University of London by Dr. PETER FAY as described by STEWART *et al.* (1967). Cultures grown in the modified GORHAM et al. (1964) medium ASM-1 with or without nitrate at 22.5 °C and 2200 lux, with orbital shaking for 10 days, were assayed under 0.1 atm acetylene for 30 min at 1500 lux, or in the dark at 22.5 °C.

Mutagen: N-methyl-N'-nitro-N-nitrosoguanidine (NTG) was a product of ALDRICH Chemical Co., Milwaukee, Wisconsin, and was given to us by Dr. C. VAN BAALEN. Fresh solutions of NTG were prepared aseptically in sterile nitrogen-free medium.

Results

Two phenotypic classes of morphological mutants were isolated. One class of mutant (*Nl 'A' mam*) shows abnormally larger cells with impaired growth rate and increased heterocyst frequency. The other class (*Nl 'A' chsp*) exhibits changed heterocyst spacing pattern.

Mutant with abnormal cell size and shape

Mutant clones of *Nl 'A' mam* exhibited high sensitivity to slightly unfavourable growth conditions as, for example, high light intensity, darkness, changes in pH, *etc.* Most of the mutant clones of this class either reverted to wild type or failed to survive in the medium. It was found that all mutant clones had very slow growth rates as compared to the parent strain.

Morphologically aberrant forms of all three cell types, viz., vegetative cells, heterocysts, and spores were observed (Figs. 1a—1g). The abnormal cells were unusually large and exhibited diverse shapes. Significant differences in the dimensions of vegetative cells, heterocysts and spores have been recorded (Table 1). Abnormal

Table 1
Cell dimensions of parent and mutant clone (*Nl mam* 101)

Cell type	Length					Width				
	Mean length (µm)		Difference	Standard error of difference	Probability x/σ (tabular value)	Mean width (µm)		Difference	Standard error of difference	Probability x/σ (tabular value)
	Parent	Mutant				Parent	Mutant			
Vegetative cells	5.0	9.0	4.0	0.1225	0.00001	4.0	5.40	1.40	0.08878	0.00001
Heterocysts	5.40	8.10	2.70	0.0763	0.00001	5.10	6.95	1.85	0.0833	0.0000374
Spores	6.90	10.3	3.40	0.1029	0.00001	6.30	9.80	3.50	0.14632	0.00001

spores were deep bluish green, thick-walled, and quite large in size with dense granular contents (Fig. 1e und 1f). Several abnormal spores were found to have lost their contents and viability (Fig. 1g). The frequency of these abnormal cells was recorded. Examination of the culture revealed a heterogeneous diversity of normal and abnormal cells (Table 2). This feature appeared to be a stable trait which was regularly observed in the progeny of single spores.

The mutant clone *NL 'A' mam* 122 showed the maximum frequency of abnormal cells and it was also found to be the most sluggish in growth.

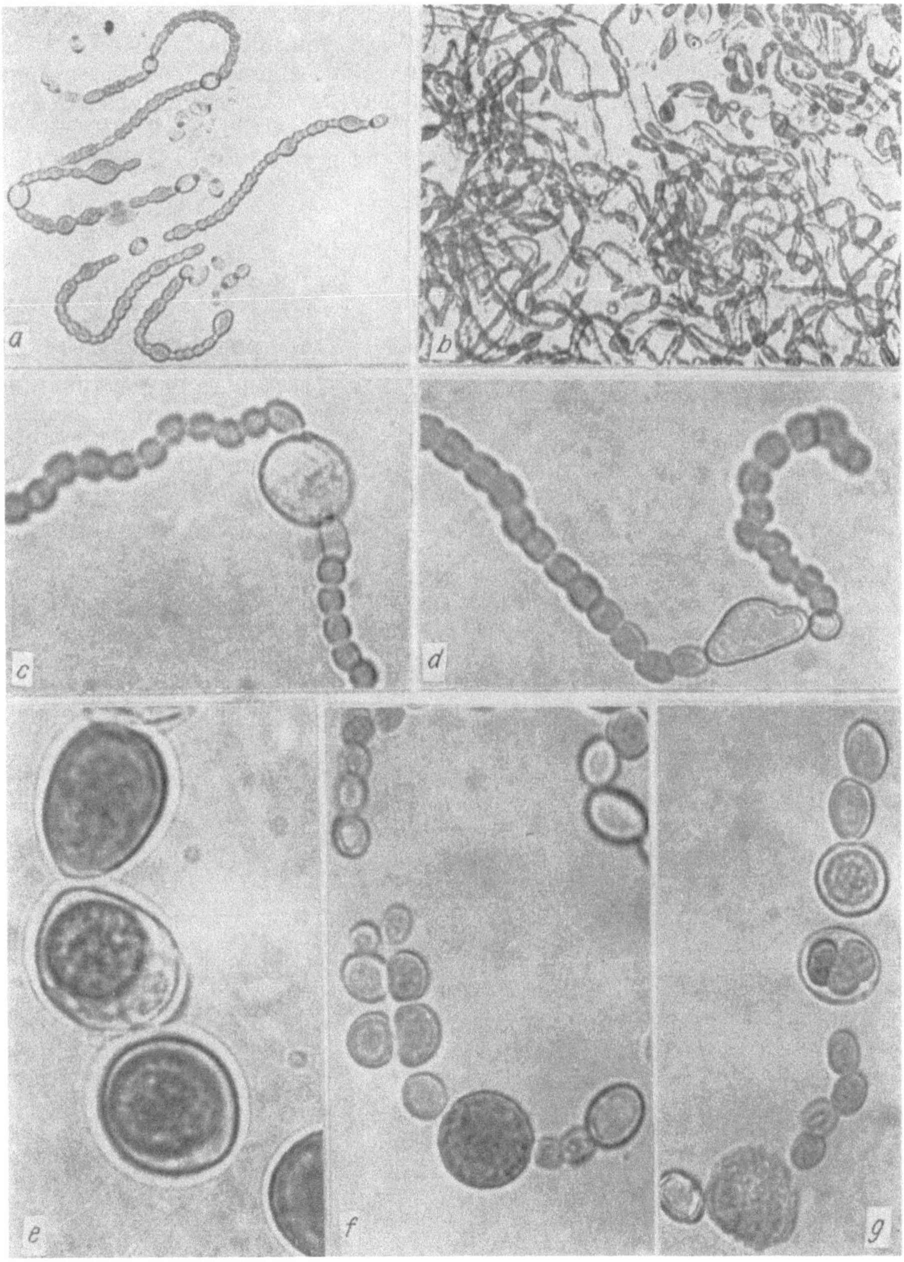

Fig. 1. a) Photomicrograph of heterocystous filaments (grown in nitrogen-free medium) showing abnormal vegetative cells, × 600; b) photomicrograph of nonheterocystous filaments (grown in ammonium medium) showing a population of abnormal vegetative cells, × 250 ; c) and d) abnormal heterocysts, × 1550 and × 1050; e) and f) abnormally enlarged and thick-walled spores, × 2750 and × 1750; g) an abnormal degenerating spore, × 1750

The frequency of abnormal cells was higher in the medium supplemented with 1 mM ammonium chloride (Table 2). However, ammonium chloride also causes lethality after a period of four days. The frequency of abnormal cells increased after 2—3 days of inoculation in medium containing 1 mM ammonium chloride. Later these abnormal cells died or lysed in the medium and only the normal cells survived and grew further in the presence of ammonium chloride. The heterocysts of the mutants showed repressibility with ammonium chloride and in this respect resembled those of the parental strain.

Table 2

Frequency of normal and abnormal cells in parent and mutant clones (*Nl mam*). a = Frequency of total cells (%), b = frequency of normal cells (%), c = frequency of abnormal cells (%)

Cell types and source of nitrogen	*Nl* '*A*' Parent			*Nl* '*A*' *mam* 101			*Nl* '*A*' *mam* 108			*Nl* '*A*' *mam* 122		
	a	b	c	a	b	c	a	b	c	a	b	c
Nitrogen free medium												
(After 7 days)												
Vegetative cells	100	100	—	100	65	35	100	58	42	100	40	60
Heterocysts	6	6	—	11	6	6	10	4	6	14	4.5	9.5
(After 21 days)												
Spores	35	35	—	65	45	20	60	45	15	75	30	45
With 1 mM NH_4Cl												
(After 3 days)												
Vegetative cells	100	100	—	100	30	70	100	40	60	100	20	80
Heterocysts	4.5	4.5	—	9	7	2	8.5	3	5.5	12.5	3	9.5

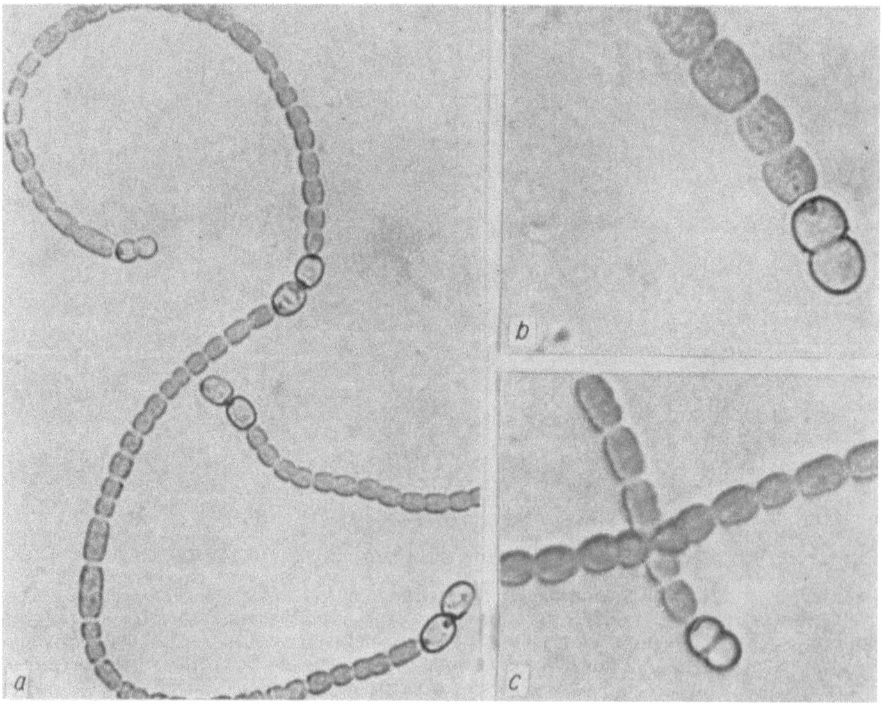

Fig. 2. a) Photomicrograph of filaments of the mutant *Nl* '*A*' *chsp* showing paired heterocysts, × 750; b) and c) dividing heterocysts in mutant *Nl* '*A*' *chsp*, × 950 and 1050

Mutants with changed heterocyst spacing pattern

In this class of mutants (*Nl 'A' chsp*) the spacing pattern of heterocysts became altered and the inter-heterocyst distance decreased. Chains of 2—8 heterocysts differentiated in series, here and there, though the majority of heterocysts occurred in pairs (Fig. 2a). Filament lengths were also decreased compared with vegetative cells (Fig. 3).

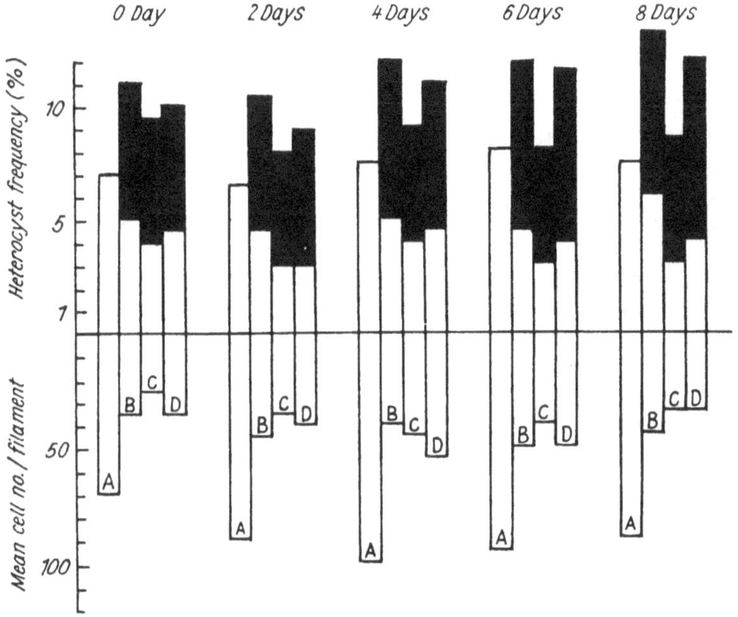

Fig. 3. Histograms showing periodic heterocyst frequency and corresponding mean cell No. per filament of (A) parent *N. linckia*, (B) *Nl 'A' chsp* 201, (C) *Nl 'A' chsp* 208, and (D) *Nl 'A' chsp* 214; upper black portion of histograms shows frequency of multiple heterocysts and lower portion shows frequency of single heterocysts

Occasionally, we have also observed dividing heterocysts. In these, the constriction could be clearly seen (Figs. 2b, 2c). Several stages from the initial constriction of the heterocyst wall to a complete constriction were observed in some cases.

Acetylene reduction by different strains of *N. linckia*

Table 3 shows the acetylene-reduction data for different strains of *N. linckia* in aerobic (in the light and in the dark) or microaerobic conditions with and without

Table 3
Acetylene reduction by parent and different mutant strains of *N. linckia*

Strain	Nitrogen source	nmoles C_2H_4 produced/mg dry alga/hr		
		Under 1% CO_2 in air		Under 1% CO_2 in N_2
		Light	Dark	Light
Parent	N_2	32.7	32.0	57.5
	NO_3^-	0.0	0.1	0.2
Nl chsp 201	N_2	7.0	5.5	20.8
	NO_3^-	0.0	0.0	0.4
Nl mam 101	N_2	11.5	7.8	25.0
	NO_3^-	0.0	0.0	0.1

nitrate. Nitrate in the medium almost completely inhibited nitrogenase activity in all the strains. Nitrogenase activity of both the mutant strains was reduced to almost one half.

Discussion

INGRAM and VAN BAALEN (1970) induced filamentous mutants in the coccoid blue-green alga *Agemenellum quadruplicatum* strain BGl after treatment with NTG. These mutants were stable and did not revert to the normal cell size and shape when grown under different growth conditions or upon the addition of various potential reversing agents to the basal medium. Growth rates of these mutants were similar to that of the wild-type. In our study, most of the variant clones of the first class of mutants either reverted back to the parental condition or died when grown under unfavourable conditions. They also showed decreased growth rates as compared to the parent. These mutants are phenotypically very similar to those reported by SINHA and KUMAR (1973) in *Anabaena doliolum.* They observed these abnormal structures when the alga was grown in a medium containing nitrate or ammonium, but in the present study these bizarre forms were also found to be produced in nitrogen-free medium. They observed a very low frequency of these abnormal cells, approximately similar to the heterocyst frequency, and they assumed that these abnormal cells might possibly represent such vegetative cells as had failed to differentiate in a medium containing ammonium or nitrate. However, we do not find any support for this latter view since the frequency of aberrant cells in these mutants is much greater than the heterocyst frequency.

Nothing is known about the cause of occurrence of these abnormal cells. It seems likely that these cells might have lost the function of division, which may cause the accumulation of metabolic products resulting in the formation of abnormal, bizarre forms.

Heterocysts occur at characteristic intervals along the length of filaments and these intervals constitute the normal spacing pattern of heterocysts. WILCOX *et al.* (1973) and WOLK (1975) suggest that a heterocyst inhibitory diffusible substance produced within the developing heterocysts and travelling in a gradient fashion along both sides is the mechanism which maintains the normal heterocyst spacing pattern. On the basis of this finding and other results with several classes of non-nitrogen fixing mutants in *N. linckia* and *N. muscorum,* we propose that the regulation of heterocyst formation is governed by a combination of a positive (inducer) and a negative (inhibitor) control mechanism. The inducer is required for the differentiation of heterocyst while the inhibitor maintains the spacing pattern (SINGH *et al.*, unpublished). Therefore, it seems that the second class of mutant (*Nl 'A' chsp*) might be partially or fully defective in the formation of the inhibitor, thus leading to the abnormal heterocyst spacing pattern.

LADHA and KUMAR (1975) have reported the division of heterocysts in two filamentous blue-green algae isolated from nature. Here we reconfirm the same phenomenon in a NTG-induced mutant of *N. linckia, Nl 'A' chsp.* The observation of a dividing heterocyst may not be a case of sudden conversion of a dividing vegetative cell to a heterocyst because heterocyst differentiation from a vegetative cell comprises many distinct developmental stages occurring chronologically. Such division also does not seem to be a case of regression or dedifferentiation of heterocyst to vegetative cell, since we believe that a mature heterocyst cannot dedifferentiate into a vegetative cell.

Our mutant strains producing unusually frequent heterocysts with no concomitant increase in nitrogenase activity strongly suggest that active nitrogenase is not obligatory for heterocyst differentiation. This is further supported by our mutants of *N. linckia* (LADHA and KUMAR, unpublished) and that of *N. muscorum* (STEWART and SINGH 1975) which does not fix nitrogen but still produces heterocysts.

Acknowledgements

We thank Dr. H. N. SINGH for comments. JKL gratefully acknowledges the receipt of a Senior Research Fellowship of the Department of Atomic Energy, Bombay.

References

GORHAM, P. R., MCLACHLAN, J. S., HAMMER, U. T. and KIM, W. K., 1964. Isolation and culture of toxic strains of *Anabaena flos-aquae* (LYNG.) DE BREB. Verh. int. Ver. theor. angew. Limnol., **15**, 796–804.

GREENBERG, J., MANDELL, J. D. and WOODY, P. L., 1961. A preliminary report — Resistance and cross-resistance of *E. coli* mutants to radiomimetic agents. Cancer Chemother. Reports, No. **11**, 51–56.

INGRAM, L. O. and VAN BAALEN, C., 1970. Characteristics of a stable, filamentous mutant of a coccoid blue-green alga. J. Bacteriol., **102**, 784–789.

KUMAR, H. D., 1963. Effects of radiations on blue-green algae. I. The production and characterization of a strain of *Anacystis nidulans* resistant to ultraviolet radiation. Ann. Bot., N.S. **27**, 723–733.

KUMAR, H. D., 1968. Action of mutagenic chemicals on *Anacystis nidulans*. III. N-Methyl-N′-nitro-N-nitrosoguanidine. Arch. Mikrobiol., **63**, 95–102.

KUMAR, H. D. and LADHA, J. K. 1976. Comparative studies of growth and nitrogen-fixation in certain subclones of the cyanophyte *Gloeocapsa*. Biochem. Physiol., Pflanzen, **169**, 64–69.

LADHA, J. K. and KUMAR, H. D., 1975a. Developmental stages in a non-heterocystous filamentous cyanophyte. New Phytol., **74**, 477–483.

LADHA, J. K. and KUMAR, H. D., 1975b. Heterocyst division in two blue-green algae. Arch. Mikrobiol., **102**, 171–173.

SINHA, B. D. and KUMAR, H. D., 1973. A mutational study of the nitrogen-fixing blue-green alga *Anabaena doliolum*. Ann. Bot., N S. **37**, 673–679.

STEWART, W. D. P., FITZGERALD, G. P. and BURRIS, R. H., 1967. *In situ* studies on nitrogen fixation using the acetylene reduction technique. Proc. nat. Acad. Sci. USA, **58**, 2071–2078.

STEWART, W. D. P. and SINGH, H. N., 1975. Transfer of nitrogen-fixing (NIF) genes in the blue-green alga *Nostoc muscorum*. Biochem. Biophysic Res. Commun. **62**, 62–69.

VAN BAALEN, C., 1975. Mutation of the blue-green alga, *Anacystis nidulans*. Science, **149**, 70.

VAN BAALEN, C., 1973. Mutagenesis and genetic recombination. In: CARR, N. G. and WHITTON, B. A. (Editors), The Biology of Blue-green Algae. Blackwell Oxford.

WILCOX, M., MITCHISON, G. J. and SMITH, Q. J., 1973. Pattern formation in the blue-green alga *Anabaena*. II. Controlled proheterocyst regression. J. Cell. Science, **13**, 637–649.

WOLK, C. P., 1975. Differentitation and pattern formation in filamentous blue-green algae. In: Spores VI. American Society for Microbiology.

Mailing address: Prof. Dr. H. D. KUMAR,
Department of Botany
Banaras Hindu University
Varanasi—221005, India

Zeitschrift für Allgemeine Mikrobiologie	17	7	1977	521–529

(Institut für medizinische und allgemeine Mikrobiologie, Virologie und Epidemiologie, Abt. Biochemie der Viren und Mikroorganismen, Humboldt-Universität Berlin)

Einfluß der Temperatur auf die Transkription der T3-DNA durch T3-spezifische RNA-Polymerase in zellfreien Extrakten von *Escherichia coli* CRT266[1])

H. Musielski, W. Mann, B. Degen, S. Michel und G. Richter

(Eingegangen am 2. 8. 1976)

Cell free extracts were prepared from *E. coli* CRT266 9 min after infection with T3 phages. RNA synthesis in these extracts is almost entirely due to T3 RNA polymerase. The inactivation of T3 RNA polymerase in these extracts proceeds rapidly at 42 °C. 90% of the activity is lost within 10 min at this temperature. Under conditions where the formation of a stable initiation complex with T3 DNA is possible, *i.e.*, in the presence of GTP, ATP, and UTP the T3 RNA polymerase becomes protected against heat inactivation losing only 10% of its activity during an exposure to 42 °C for 10 min. Studies on the time course of RNA synthesis have shown that reinitiation is still possible at 37 °C and 42 °C. At 44 °C, however, RNA synthesis stops abruptedly after 3 min indicating that reinitiation does no longer take place. The elongation of already initiated T3 RNA chains is rather resistant to heat. At 44 °C the same elongation rates are observed as at 37 °C and 42 °C, respectively.

E. coli CRT266, eine *ts*-Mutante mit einer Läsion im *dnaB*-Gen, zeigt nach Infektion mit T3-Phagen bei der nonpermissiven Temperatur von 41,5 °C weder Lyse des Wirtsbakteriums noch T3-DNA-Synthese noch Bildung von Lysozym (Mann *et al.* 1972, Elgammal 1973). Diese Befunde weisen darauf hin, daß die Transkription der späten Gene des Phagen T3 durch die im Gen 1 kodierte T3-spezifische RNA-Polymerase bei 41,5 °C nicht stattfindet. Mittels toluolbehandelter Bakterien konnte diese Vermutung bestätigt werden (Mann *et al.* 1972). Da bei der nonpermissiven Temperatur die T3-spezifische RNA-Polymerase synthetisiert wird (Mann *et al.* 1972), kann das Ausbleiben der Transkription der späten Gene sowohl *in vivo* als auch in der toluolbehandelten Mutante nicht auf die Abwesenheit des genannten Enzyms zurückgeführt werden. Es erschien daher wichtig, den Einfluß der Temperatur auf die Transkription der T3-DNA durch T3-RNA-Polymerase zu untersuchen. Die hierbei erhaltenen Ergebnisse zeigen, daß die Hitzestabilität der T3-RNA-Polymerase von ihrer Bindung an die T3-DNA abhängt. In der Form eines stabilen Komplexes mit der T3-DNA ist das Enzym wesentlich resistenter gegen Hitzeinaktivierung als im ungebundenen Zustand.

Material und Methoden

Bakterien und Bakteriophagen: *E. coli* CRT266 (*thr, leu, lac*$_y$, λ^s, *thy*, F^-, *dnaB-266*) erhielten wir von Prof. Dr. F. Jacob, *E. coli* CR34-45 (*thr, leu, lac*$_y$, λ^s, *thy, strep*^{r}, T_4^r, F^-) von Prof. Dr. F. Bonhöfer. Der Bakteriophage T3 stammt aus der Sammlung von Doz. Dr. H. A. Rosenthal.

T3-DNA: Die T3-Phagen wurden nach einer früher beschriebenen Methode gereinigt (Stibenz *et al.* 1971). Zur Abtrennung der Ribosomen von den T3-Phagen wurde anstelle der DEAE-Chro-

[1]) Ein Teil der Ergebnisse wurde auf dem 9. FEBS-Kongreß in Budapest (1974) vorgetragen.

matographie eine Zentrifugation durch 20%ige Saccharose (Spinco, Rotor SW25, 20000 rpm, 4 Std.) angewandt. Aus den gereinigten T3-Phagen wurde die DNA mit Phenol nach Thomas u. Abeleson (1966) extrahiert.

Herstellung des zellfreien Extraktes: Die zellfreien Extrakte aus infizierten *E. coli* CRT266 wurden in Anlehnung an Strätling u. Knippers (1971) wie folgt präpariert: *E. coli* CRT266 bzw. *E. coli* CR34—45 wurden in $M9_Z$ (Mann *et al.* 1972) bei 30 °C gezüchtet und die logarithmisch wachsenden Bakterien (5×10^8 Zellen/ml) mit T3-Phagen infiziert (Bakterien und Phagen im Verhältnis 1:10). 9 min nach Infektion wurde Chloramphenicol (300 µg/ml) zugesetzt. Nach einer weiteren Minute wurden die Suspension auf Eis gegossen und die Bakterien anschließend abzentrifugiert (5000 g, 10 min). Das Bakteriensediment wurde 2mal mit Puffer I (0,05 M Tris-HCl, pH 7,5, 0,01 M $MgCl_2$, 0,022 M NH_4Cl und 300 µg/ml Chloramphenicol) gewaschen und danach in Puffer II (0,05 M Tris-HCl, pH 7,5, 0,005 M EDTA, 15%ige Saccharose und 250 µg/ml Lysozym) suspendiert. Pro Gramm Bakterienfeuchtgewicht wurden 10 ml Puffer II benutzt. Zur Ausbildung von Sphäroplasten wurde diese Suspension eine Stunde im Eisbad inkubiert. Zur Lyse der Sphäroplasten wurden der Suspension das neutrale Detergens Brij 58 (2 mg/ml), 0,001 M Dithioerythrit (DTE) und 0,005 M $MgSO_4$ zugesetzt, und das Gemisch wurde 10 min bei 25000 rpm im Ti50-Rotor (Spinco) zentrifugiert und der Überstand gegen 55% Glycerin in 0,05 M Tris-HCl und 0,01 M β-Mercaptoäthanol bei pH 7,8 dialysiert. Dieser zellfreie Extrakt kann bei —20 °C ohne wesentlichen Verlust an Aktivität mehrere Monate aufbewahrt werden.

Zur Herstellung von zellfreien Extrakten aus bei 42 °C infizierten *E. coli* CRT266 wurden die logarithmisch wachsenden Bakterien auf diese Temperatur erwärmt und nach einer Inkubationsperiode von 20 min mit T3-Phagen infiziert. 9 min später wurden die infizierten Bakterien wie oben beschrieben aufgearbeitet.

RNA-Synthese: 1. Standardreaktionsbedingungen: Die Aktivität der RNA-Polymerase wurde in einem Reaktionsgemisch, das folgende Komponenten pro ml Ansatz enthielt, bestimmt: 50 mM Tris-HCl (pH 7,8 bei 23 °C), 20 mM $MgCl_2$, 0,1 mM EDTA, 0,1 mM DTE, 0,6 mM ATP, 0,6 mM GTP, 0,6 mM CTP, 0,3 mM (^3H) UTP ($5,8 \times 10^4$ dpm/nMol), 0,6 mg Rinderserumalbumin, 10 µg Rifampicin, 30—40 µg T3-DNA und 40 µl zellfreien Extrakt entsprechend 120—150 µg Protein. Das Reaktionsgemisch wurde 10 min bei 37 °C inkubiert. Die RNA-Synthese wurde durch Zugabe des gleichen Volumens von 10%iger eiskalter Trichloressigsäure, die 0,02 M $Na_2P_2O_7$ enthielt, abgestoppt.

Der gebildete Niederschlag wurde auf Filterpapierscheiben (Filtrak FN8) gesammelt und mit 5%iger Trichloressigsäure, Äthanol-Äther und Äther in angegebener Reihenfolge gewaschen. Die trockenen Filterpapierscheiben wurden in Szintillationsküvetten mit 0,1 ml Soluene (Packard Co.) versetzt, und nach Zugabe einer Szintillatorlösung auf Toluolbasis wurde die Radioaktivität mit dem LKB-Wallace Scintillationszähler gemessen. Zur Untersuchung des Einflusses der Temperatur und der Nucleosidtriphosphate auf die T3-spezifische RNA-Synthese wurde das Standardreaktionsgemisch wie folgt abgeändert: Reaktionsgemisch I ist das Standardreaktionsgemisch ohne CTP und (^3H) UTP. Reaktionsgemisch II enthält alle Komponenten des Standardreaktionsgemisches außer den 4 Nucleosidtriphosphaten und (^3H) UTP. Reaktionsgemisch III entspricht Reaktionsgemisch II ohne T3-DNA. Zur Messung der Aktivität der T3-RNA-Polymerase wurden die einzelnen Reaktionsgemische mit den fehlenden Komponenten komplettiert, wobei die Endkonzentrationen aller Komponenten denen der Standardreaktionsmischung entsprachen.

2. RNA-Synthese bei verschiedenen Temperaturen: Die folgenden zwei Bedingungen wurden gewählt:

a) Vorinkubation in Gegenwart von GTP, ATP und UTP: Proben von je 0,2 ml des Reaktionsgemisches I wurden zum Temperaturausgleich von 0 °C auf 37 °C, 39 °C, 41 °C, 42 °C, 43 °C und 44 °C 2 min in getrennten Wasserbädern vorinkubiert. Die RNA-Synthese wurde durch Zugabe von CTP und (^3H) UTP gestartet. Das Endvolumen betrug 0,25 ml. 10 min nach Start der RNA-Synthese wurde die Radioaktivität in dem säureunlöslichen Niederschlag bestimmt.

b) Vorinkubation in Abwesenheit von Nucleosidtriphosphaten: Proben des Reaktionsgemisches II wurden wie unter a) beschrieben zum Temperaturausgleich bei den angegebenen Temperaturen inkubiert. Gestartet wurde durch Zugabe von GTP, ATP, UTP, CTP und (^3H) UTP.

Inaktivierung der T3-RNA-Polymerase bei 42 °C bzw. bei 37 °C: Eine Reihe von Proben, die 0,2 ml Reaktionsgemisch I bzw. Reaktionsgemisch III enthielten, wurde in Wasserbädern von 37 °C bzw. 42 °C inkubiert. Zu den angegebenen Zeiten wurden einzelne Röhrchen entnommen und sofort im Eisbad gekühlt. Nach Zugabe der fehlenden Komponenten des Standardreaktionsgemisches wurde die verbleibende Aktivität bei 37 °C wie oben beschrieben gemessen. Das Endvolumen betrug 0,25 ml.

Einfluß von Nucleosidtriphosphaten auf die Stabilität des T3-DNA-Enzym-Komplexes: Proben von 0,2 ml des Reaktionsgemisches II, die die in Tabelle 2 angegebenen Nucleosidtriphosphate enthielten, wurden zum Temperaturausgleich 2 min bei 42 °C inkubiert. 10 min nach Zugabe der fehlenden Nucleosidtriphosphate wurde die gebildete RNA bestimmt.

Zeitlicher Verlauf der RNA-Synthese bei verschiedenen Temperaturen: 5 ml des Reaktionsgemisches I wurden im Wasserbad von 37 °C, 42 °C und 44 °C 4 min lang vorinkubiert. Die RNA-Synthese wurde durch Zugabe von CTP und (^{3}H)UTP gestartet. Das Endvolumen betrug 6 ml. Die Endkonzentrationen der einzelnen Komponenten entsprechen denen des Standardreaktionsgemisches. Zu den angegebenen Zeiten wurden Proben von 0,2 ml entnommen, sofort in das 10-fache Volumen von eiskalter 5%iger Trichloressigsäure pipettiert, und die Radioaktivität des säureunlöslichen Niederschlags wurde bestimmt.

Ergebnisse

Die Bedingungen für die RNA-Synthese in zellfreien Extrakten, die 9 min nach Infektion mit T3-Phagen aus *E. coli* CRT 266 gewonnen worden waren, sind in Tabelle 1 dargestellt. Zugabe von Rifampicin, das die Initiation der Transkription durch die *E. coli*-RNA-Polymerase hemmt (Sippel u. Hartmann 1968), aber keine Wirkung auf die T3-RNA-Polymerase besitzt (Dunn *et al.* 1971, Maitra 1971), zeigt keinen wesentlichen Einfluß auf den Einbau von (^{3}H)UMP. Werden einzelne Nucleosidtriphosphate aus dem Reaktionsansatz weggelassen, so findet keine RNA-Synthese statt. Die Extrakte enthalten daher für unsere Zwecke nur unerhebliche Mengen von Nucleosidtriphosphaten.

Tabelle 1
RNA-Synthese in zellfreien Extrakten aus E. *coli* CRT 266 nach Infektion mit T3-Phagen

Bedingungen	Einbau von [^{3}H] UMP (%)
Komplettes System	100
Komplettes System + Rif. (10 μg/ml)	95
Komplettes System —CTP	< 1
Komplettes System —GTP	< 1
Komplettes System —ATP	< 1
Ohne T3-DNA	3
Ohne T3-DNA +T4-DNA	6
Ohne T3-DNA +Kalbsthymus-DNA	1

Die Versuche wurden unter Standardreaktionsbedingungen (s. Material und Methoden) durchgeführt. 100% entsprechen $2{,}5 \times 10^4$ Zpm.

T3-spezifische RNA-Polymerase transkribiert nur T3-DNA. T4-DNA und Kalbsthymus-DNA werden von diesem Enzym nicht als Matrize benutzt (Dunn *et al.* 1971, Maitra 1971). Wie aus Tabelle 1 ersichtlich ist, findet ohne den Zusatz von T3-DNA keine RNA-Synthese statt. T4-DNA und Kalbsthymus-DNA waren inaktiv. Diese Ergebnisse zeigen, daß in den aus T3-infiziertem *E. coli* CRT266 gewonnenen zellfreien Extrakten die RNA-Synthese allein durch die T3-induzierte RNA-Polymerase erfolgt. Um den Einfluß der Temperatur auf die Transkription der T3-DNA durch die T3-RNA-Polymerase zu untersuchen, wurden zwei unterschiedliche Versuchsanordnungen gewählt:

a) Die RNA-Synthese im zellfreien Extrakt wurde zusammen mit T3-DNA, GTP, ATP und UTP gestartet.

b) Die Vorinkubation bei den verschiedenen Temperaturen fand in Abwesenheit der drei genannten Nucleosidtriphosphate statt. In diesen Experimenten wurde zum Start der RNA-Synthese dem Versuchsansatz ATP, GTP, UTP, CTP und (^{3}H)UTP zugesetzt.

Die Ergebnisse sind in Abbildung 1 dargestellt. Es ist ersichtlich, daß der Einfluß der Temperatur auf die T3-spezifische RNA-Synthese stark von den Vorinkubationsbedingungen abhängig ist. Fand die Vorinkubation in Gegenwart von GTP, ATP und UTP statt, so war die RNA-Synthese im Vergleich zu den in Abwesenheit dieser Nucleosidtriphosphate erhaltenen Ergebnissen wesentlich resistenter gegenüber erhöhten Temperaturen. Bei 42 °C wurden unter diesen Inkubationsbedingungen fast die gleichen Werte gefunden wie bei 37 °C, und bei 44 °C sank dieser Wert etwa auf die Hälfte. Wird dagegen die Vorinkubation in Abwesenheit der genannten drei Nucleosidtriphosphate durchgeführt, so beobachtet man bei 42 °C einen im Vergleich zu 37 °C um 85% erniedrigten (^{3}H)UMP-Einbau. Bei 44 °C kam dieser Prozeß vollkommen zum Erliegen.

Um den Einfluß der Temperatur auf die T3-spezifische Transkription genauer zu charakterisieren, wurde die Inaktivierung der T3-RNA-Polymerase bei 42 °C unter

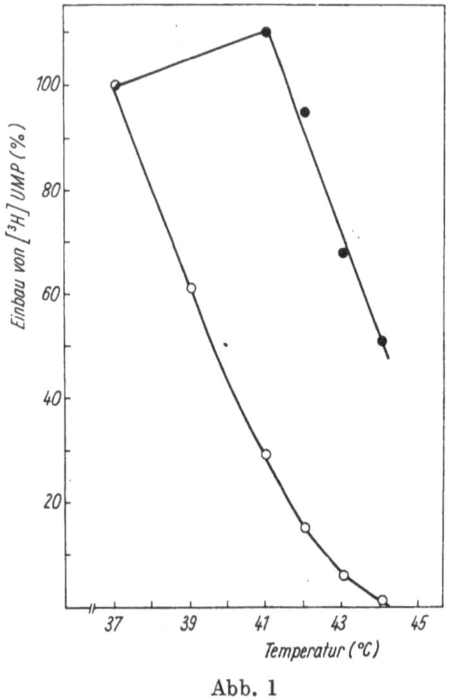

Abb. 1

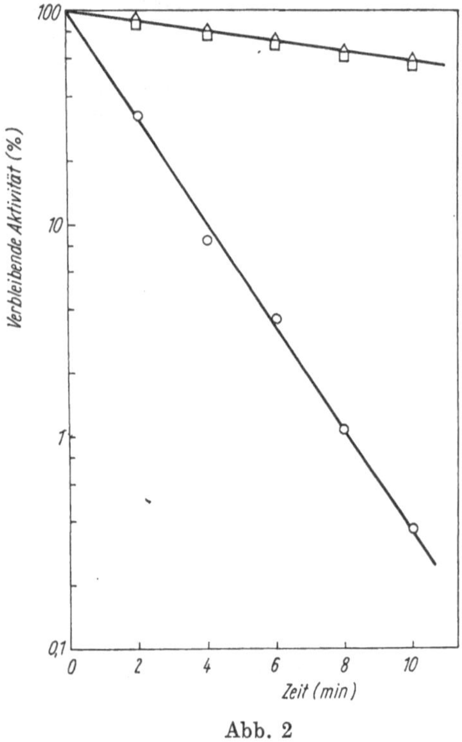

Abb. 2

Abb. 1. Einfluß der Temperatur auf die T3-spezifische RNA-Synthese. ● Bedingung a): Temperaturausgleich fand in Gegenwart von ATP, GTP, UTP und T3-DNA statt. Die RNA-Synthese wurde durch Zugabe von CTP und (^{3}H) UTP gestartet

○ Bedingung b): Temperaturausgleich erfolgt in Abwesenheit von Nucleosidtriphosphaten. Die RNA-Synthese wurde durch Zugabe von ATP, GTP, UTP, CTP und (^{3}H) UTP gestartet. 100% entsprechen 22,5 × 10^4 Zpm. Einzelheiten siehe Material und Methoden

Abb. 2. Inaktivierung der T3-RNA-Polymerase in zellfreien Extrakten bei den Temperaturen 37 °C und 42 °C.

△: Abfall der Enzymaktivität bei 42 °C in Anwesenheit von ATP, GTP, UTP und T3-DNA.

Inaktivierung des Enzyms bei 37 °C (□) bzw. 42 °C (○) in Abwesenheit von Nucleosidtriphosphaten und T3-DNA. Die Versuchsbedingungen sind in Material und Methoden angegeben. 100% entsprechen 30 × 10^4 Zpm den beiden oben beschriebenen Bedingungen untersucht. Die hierbei erhaltenen Ergebnisse sind in Abbildung 2 dargestellt. Nach Vorinkubation in Gegenwart von T3-DNA, GTP, ATP und UTP (Bedingung a) verliert das Enzym bei 42 °C nur langsam an Aktivität. Die in diesen Versuchen beobachtete Inaktivierungsrate entspricht dem Wert, der bei 37 °C für den zellfreien Extrakt allein (Bedingung b) gemessen wurde. Wird dagegen die Inkubation bei 42 °C in Abwesenheit von GTP, ATP, UTP und T3-DNA durchgeführt (Bedingung b), so kommt es zu einem schnellen Absinken der T3-RNA-Polymerase-Aktivität in den zellfreien Extrakten.

Zugabe von T3-DNA ohne die genannten drei Nucleosidtriphosphate schützt das Enzym nicht vor Inaktivierung bei 42 °C (diese Ergebnisse sind in der Abb. 2 nicht dargestellt).

Wie aus den Arbeiten von McAllister *et al.* (1973) und Küpper *et al.* (1973) bekannt ist, wird in Gegenwart von GTP, ATP und UTP ein stabiler Initiationskomplex zwischen T3-DNA und T3-RNA- Polymerase ausgebildet, der gegenüber Rifampicin AF 013 und Heparin resistent ist. Es erschien daher möglich, daß die unter Bedingung a) beobachtete Hitzeresistenz auf die Ausbildung eines Initiationskomplexes zwischen T3-DNA und T3-RNA Polymerase während des Temperaturausgleichs zurückzuführen ist. Daher wurde der Einfluß der Vorinkubation mit GTP, ATP, UTP und CTP einzeln und in verschiedenen Kombinationen auf die RNA-Synthese bei 42 °C untersucht. Die Ergebnisse dieser Versuche sind in Tabelle 2 wiedergegeben. Der höchste Einbau von (^{3}H)UMP in die säurelösliche Fraktion fand nach Vorinkubation in Gegenwart von GTP, ATP und UTP statt. Der unter diesen Bedingungen gefundene Wert wurde gleich 100% gesetzt. Von den einzelnen Nucleosidtriphosphaten zeigte allein GTP eine Wirkung. Zusatz von ATP zu GTP erhöhte die Hitzestabilität der RNA-Synthese weiter, während CTP und UTP diesen Effekt nicht zeigen. Eine zusätzliche Steigerung der Thermostabilität der T3-spezifischen Transkription konnte nach Zugabe von UTP zu dem Gemisch von GTP und ATP erzielt werden.

Tabelle 2
Einfluß von Nucleosidtriphosphaten auf die Stabilität des DNA-Enzym-Komplexes bei 42 °C

Nucleosidtriphosphate	Einbau von [^{3}H] UMP (%)
ATP	15
GTP	35
CTP	17
UTP	16
ATP, GTP	86
ATP, CTP	14
ATP, UTP	13
GTP, CTP	35
GTP, UTP	41
UTP, CTP	14
ATP, GTP, CTP	88
ATP, GTP, UTP	100
ATP, CTP, UTP	12
GTP, CTP, UTP	37

Die in der Tabelle angegebenen Nucleosidtriphosphate waren während des Temperaturausgleiches auf die Temperatur 42 °C im Reaktionsgemisch vorhanden.
Einzelheiten der Versuchsdurchführung siehe Material und Methoden. 100% entsprechen 22,5 × 10^4 Zpm

Wie aus Abbildung 1 ersichtlich ist, wird nach Vorinkubation mit GTP, ATP und UTP die T3-spezifische RNA-Synthese bei 44 °C im Vergleich zu 37 °C um ungefähr 50% reduziert. Durch die Arbeiten von Maitra u. Huang (1972), McAllister *et al.* (1973) und Küpper *et al.* (1973) ist bekannt, daß für die Transkription des T3-Genoms *in vitro* 2—4 min benötigt werden. In unseren Versuchen wurde der (^{3}H)UMP-Einbau nach einer Inkubationszeit von 10 min gemessen. Als Ursache für die Hemmung der RNA-Synthese bei 44 °C muß daher eine Erniedrigung der Elongationsrate schon initiierter T3-RNA-Ketten oder ein Ausbleiben der Reinitiation neuer T3-RNA- Ketten in Betracht gezogen werden.

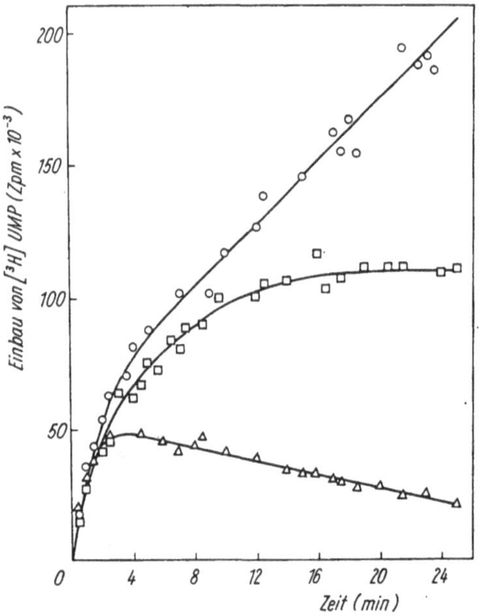

Abb. 3. Zeitlicher Verlauf der RNA-Synthese bei verschiedenen Temperaturen. ○ 37 °C, □ 42 °C, △ 44 °C.
Die Versuchsbedingungen sind in Material und Methoden angegeben

Um zwischen diesen beiden Möglichkeiten zu unterscheiden, wurde der zeitliche Verlauf der T3-spezifischen RNA-Synthese bei 37 °C, 42 °C und 44 °C verfolgt. Abbildung 3 zeigt die Ergebnisse dieser Versuche. Bei 37 °C läuft die RNA-Synthese über den gesamten untersuchten Zeitraum von 25 min ab. Bei 42 °C verlangsamt sich die Synthesegeschwindigkeit nach ungefähr 4 min und kommt nach 15 min zum Erliegen, während bei 44 °C die T3-RNA-Synthese anfänglich mit derselben Geschwindigkeit wie bei den anderen beiden Temperaturen abläuft, jedoch zwischen der 3. und 4. Minute abrupt abstoppt. Das entspricht der Zeit, die für die Transkription der T3-DNA *in vitro* benötigt wird (Maitra u. Huang 1972, McAllister *et al.* 1973, Küpper *et al.* 1973).

Diskussion

Unsere Untersuchungen haben gezeigt, daß in zellfreien Extrakten, die aus *E. coli* CRT266 bzw. *E. coli* CR34—45 9 min nach Infektion mit T3-Phagen hergestellt worden waren, die Transkription der T3-DNA fast ausschließlich von der T3-induzierten RNA-Polymerase durchgeführt wird. Möglicherweise wird ein Teil der *E. coli*-RNA-Polymerase zusammen mit der Wirts-DNA bei der Herstellung der Extrakte entfernt. Da jedoch schon früh während des Replikationszyklus des Bakteriophagen T3 die Wirts-DNA abgebaut wird (HAUSMANN u. GOMEZ 1971, 1972), erscheint es durchaus möglich, daß die Extrakte *E. coli*-RNA-Polymerase enthalten. Durch die Arbeiten von MAHADIK *et al.* (1972, 1974) ist bekannt, daß nach Infektion mit T3-Phagen ein Hemmstoff für die *E. coli*-RNA-Polymerase gebildet wird. Das Gen, das diesen Inhibitor codiert, gehört zu den „early late" Genen (MAHADIK *et al.* 1974) und wird von der T3-spezifischen RNA-Polymerase transkribiert. Unter unseren Bedingungen, d. h. 9 min nach Infektion, müßte dieser Hemmstoff durchaus schon synthetisiert sein. Ein weiterer Faktor, der eine Rolle spielen könnte, wurde ebenfalls von diesen Autoren beschrieben (DHARMGRORGARTAMA *et al.* 1973). Sie fanden, daß es nach Infektion mit T3-Phagen zu einer Alteration der β'-Untereinheit der *E. coli*-RNA-Polymerase kommt, und daß das so veränderte Enzym nur noch ungefähr ein Viertel der Aktivität der normalen *E. coli*-RNA-Polymerase besitzt. Aus den genannten Gründen ist zu erwarten, daß die Wirts-RNA-Polymerase, sollte sie in den zellfreien Extrakten vorhanden sein, weitgehend inaktiv ist.

Weiterhin zeigen unsere Untersuchungen, daß unter Bedingungen, unter denen ein stabiler Initiationskomplex zwischen T3-DNA und T3-RNA-Polymerase gebildet wird, das Enzym gegen Inaktivierung bei 42 °C geschützt ist. Während das freie, nicht DNA-gebundene Enzym bei der genannten Temperatur innerhalb von 10 min mehr als 90% seiner Aktivität verliert, beträgt der Verlust nach Ausbildung des Inititationskomplexes weniger als 10%. Dies bedeutet, daß die Bindung der T3-RNA-Polymerase an die T3-DNA die Stabilität des Enzyms gegen Hitzeinaktivierung stark erhöht. Zugabe von T3-DNA allein ist ohne Einfluß auf die Thermostabilität der T3-RNA-Polymerase. Dies zeigt, daß im Unterschied zu der *E. coli*-RNA-Polymerase, die in Gegenwart von DNA noch bei 50 °C stabil ist (STEAD u. JONES 1967), die Bindung der T3-DNA-Polymerase an die T3-DNA nur sehr schwach sein kann.

Vorinkubation der T3-RNA-Polymerase mit T3-DNA und GTP bzw. mit T3-DNA und einem Gemisch aus GTP und ATP oder der Kombination von GTP, ATP und UTP erhöht in der angebenen Reihenfolge die Stabilität der T3-spezifischen RNA-Synthese bei 42 °C (Tab. 2). Diese Ergebnisse weisen darauf hin, daß zur Ausbildung eines thermoresistenten Komplexes zwischen T3-DNA und T3-RNA-Polymerase ein Oligonucleotid von der Sequenz $ppp(Gp)_m(Ap)_n(U)_o$ (m, n, o $\geqq$ 1) synthetisiert wird. Dieses Resultat stimmt mit den Befunden von McALLISTER *et al.* (1973) und MAITRA *et al.* (1974) überein, die berichteten, daß zur Bildung eines Initiationskomplexes die Synthese des Oligonucleotids $pppGp(Gp)_n(Ap)_m$ (n, m $\geqq$ 1) notwendig ist. Weiterhin zeigen unsere Untersuchungen, daß sich die T3-RNA-Polymerase gegenüber Inaktivierung durch Hitze ähnlich verhält wie gegenüber Inaktivierung durch Heparin (McALLISTER *et al.* 1973) oder Rifampicin AF 013 (KÜPPER *et al.* 1973).

Wie die Kinetik der Transkription der T3-DNA durch die T3-RNA-Polymerase zeigt (Abb. 3), verläuft die RNA-Synthese bei 37 °C über mehr als 25 min. Da für die *in vitro*-Transkription der T3-DNA eine Zeitspanne von 2—5 min benötigt wird (MAITRA u. HUANG 1972, McALLISTER *et al.* 1973, KÜPPER *et al.* 1973), bedeutet dies, daß bei 37 °C eine Reinitiation neuer RNA-Ketten stattfindet. Einen ganz anderen Verlauf nimmt die T3-RNA-Synthese bei 44 °C. Bei dieser Temperatur findet die Elongation schon initiierter T3-RNA-Ketten statt. Außerdem zeigt sich, daß das an die T3-DNA gebundene Enzym gegen Inaktivierung durch Hitze verhältnismäßig geschützt ist. Besonders hervorzuheben ist, daß bei 44 °C die T3-spezifische RNA-Synthese 3—4 min nach Start fast schlagartig zum Erliegen kommt, d. h. bei 44 °C scheint eine Reinitiation neuer T3-RNA-Ketten nicht mehr möglich zu sein. Wir möchten diese Ergebnisse wie folgt interpretieren: 1) Nach Ablauf der Transkription der T3-DNA wird die T3-RNA-Polymerase von der Matrize abgelöst, und das freigesetzte Enzym wird bei 44 °C so schnell inaktiviert, daß eine Reinitiation neuer T3-RNA-Ketten nicht mehr möglich ist. 2) Als ein weiterer Grund für das Ausbleiben einer Reinitiation bei 44 °C muß die Beeinflussung der Initiationsstellen (Promotorregionen) der T3-DNA durch die erhöhte Temperatur in Betracht gezogen werden. Neuere Untersuchungen in unserem Laboratorium haben gezeigt, daß die optimale Temperatur für die Ausbildung eines stabilen Intitiationskomplexes zwischen T3-DNA und T3-RNA-Polymerase bei 30 °C liegt (unveröffentlichte Ergebnisse). Bei 44 °C müßte daher die Konformation der Initiationsstellen auf der T3-DNA derartig verändert sein, daß eine Bindung der T3-RNA-Polymerase an die Promotorregionen stark erschwert ist. Bei der Temperatur von 42 °C dürften diese beiden Effekte nicht ganz so ausgeprägt sein, da die T3-spezifische RNA-Synthese über einen Zeitraum von etwa 12 min ablaufen kann. Neben den oben erwähnten Extrakten aus *E. coli* CRT266 (Infektion mit T3-Phagen bei 30 °C) wurden Extrakte aus denselben Bakterien nach Infektion bei 42 °C und Extrakte aus dem Wildtyp *E. coli* CR 34—45, infiziert bei 30 °C, hergestellt. Der Einfluß der Temperatur auf die T3-RNA-Synthese war bei allen drei Präparaten der gleiche (Ergebnisse nicht gesondert dargestellt).

Dies weist daraufhin, daß nach Infektion von *E. coli* CRT266 mit T3-Phagen bei 42 °C keine alterierte T3-RNA-Polymerase gebildet wird.

Literatur

Bautz, E. K. F., 1973. Initiation of transcription by RNA polymerase of *Escherichia coli* and phage T3. FEBS Letters, **36**, 121.

Dharmgrongartama, B., Mahadik, S. P. and Srinivasan, P. R., 1973. Modification of RNA polymerase after T3 phage infection of *Escherichia coli* B. Proc. nat. Acad. Sci. USA, **70**, 2845.

Dunn, J. J., Bautz, F. A. and Bautz, E. K. F., 1971. Different template specificities of phage T3 and T7 RNA polymerase. Nature, New Biology (London), **230**, 94.

Elgammal, S., 1973. Untersuchung zur Replikation des Phagen T3 in der temperatursensitiven Mutante *Escherichia coli* CRT266. Dissertation Humboldt-Universität Berlin.

Hausmann, R. and Gomez, B., 1967. Amber mutants of bacteriophage T3 and T7 defective in phage-directed deoxyribonucleic acid synthesis. J. Virol, **1**, 779.

Hausmann, R. and Gomez, B., 1968. Bacteriophage T3- and T7-directed deoxyribonucleases. J. Virol. **2**, 265.

Küpper, H. A., McAllister, W. T. and Bautz, E. K. F., 1973. Comparison of *Escherichia coli* and T3 RNA polymerase. Differential inhibition of transcription by various drugs. Eur. J. Biochem., **38**, 581.

Mahadik, S. P., Dharmgrongartama, B. and Srinivasan, P. R., 1972. An inhibitory protein of *Escherichia coli* RNA polymerase in bacteriophage T3-infected cells. Proc. nat. Acad. Sci. USA., **69**, 162.

Mahadik, S. P., Dharmgrongartama, B. and Srinivasan, P. R., 1974. Regulation of host ribonucleic acid synthesis in T3-infected cells. Properties of an inhibitory protein of *Escherichia coli* ribonucleic acid polymerase. J. biol. Chemistry, **249**, 1787.

Maitra, U., 1971. Induction of a new RNA polymerase in *Escherichia coli* infected with bacteriophage T3. Biochem. Biophysic. Res. Commun., **43**, 443.

Maitra, U. and Huang, H. H., 1972. Initiation, release, and reinitiation of RNA chains by bacteriophage T3-induced RNA polymerase from T3 DNA templates. Proc. nat. Acad. Sci. USA, **69**, 55.

Maitra, U., Saloo, R. A. and Chakraborty, P. R., 1974. Specificity of ribonucleic acid chain initiation by bacteriophage T3-induced ribonucleic acid polymerase. J. biol. Chemistry, **249**, 5835.

Mann, W., Helm, R., Jüngling, S. und Richter, G., 1972. Untersuchungen an T3-Phagen. II. Unterschiede in der Replikation der Phagen T3 und T7 in einer Mutante von *Escherichia coli* mit temperatursensitiver DNS-Synthese. Acta biol. med. german., **29**, 471.

Mann, W., Michel, S., Brux, B., Elgammal, S. and Richter, G., 1972. Replication of T3-phages in a temperature-sensitive mutant of *Escherichia coli* (*E. coli* CRT266). studia biophysica, 31/32, 15.

McAllister, W. T., Küper, H. and Bautz, E. K. F., 1973. Kinetics of transcription by bacteriophage T3 RNA polymerase *in vitro*. Eur. J. Biochem., **34**, 489.
Sippel, A. and Hartmann, G., 1968. Mode of action of rifampicin on RNA polymerase reaction. Biochim. biophysica Acta, **157**, 218.
Stead, N. W. and Jones, Jr. O., 1967. The binding of RNA polymerase to DNA: Stabiliziation by nucleoside triphophates. Biochim. biophysica Acta, **145**, 679.
Stibens, D., Mann, W., Brandt, B., Michel, S. und Richter, G., 1971. Präparation gereinigter T3-Phagen und gelektrophoretische Auftrennung ihrer Hüllproteine. Acta biol. med. german., **27**, 553.
Strätling, W. and Knippers, R., 1971. Deoxyribonucleic acid synthesis in isolated DNA-membrane complexes. J. molecular Biol., **61**, 471.
Thomas, Jr., C. A. and Abelson, J., 1966. In: Procedures in Nucleic Acid Research. Harper and Row Publ. New York—London.

Anschrift: Dr. H. Musielski
Bereich Medizin (Charité) der Humboldt-Universität, Institut für medizinische Mikrobiologie, Abt. Biochemie
DDR 108 Berlin, Clara-Zetkin-Str. 96

Zeitschrift für Allgemeine Mikrobiologie	17	7	1977	531–542

(USSR Academy of Sciences, Moscow, Institute of Microbiology; Akademie der Wissenschaften der DDR, Forschungszentrum für Molekularbiologie und Medizin, Zentralinstitut für Mikrobiologie und experimentelle Therapie, Jena, Direktor: Prof. Dr. U. TAUBENECK)

A mathematical model of microbial growth including an intermediate

I. Growth in batch cultures

T. A. PETROVA, W. A. KNORRE R. GUTHKE and F. BERGTER

(Eingegangen am 23. 7. 1976)

A mathematical model of microbial growth is presented and examined which, in contrast to the well-known MONOD model, includes transitions from one cell "bottle-neck" to another. This is achieved by introducing an intermediate product in the model. Three variants of the model for different regulatory functions of the intermediate are considered. The results permit to describe a set of experimentally observable microbial growth curves. According to the model, the shape of the growth curves, the kinetics of substrate consumption and changes of intermediate concentration depend on culture prehistory and the nature of the intermediate regulatory function.

The growth of a homogeneous culture of microorganisms with a single limiting substrate is generally described by the well-known MONOD model. If the inhibitory effect of a metabolic product is to be considered, then the model must take into account product formation. Three distinct cases of product synthesis are recognizable, represented by the following flow equations:

$$1.\ \alpha S \rightarrow X + \beta P;$$

$$2.\ \alpha_1 S \rightarrow X + \beta P; \qquad \alpha_2 P \rightarrow X;$$

$$3.\ \alpha_1 S \rightarrow P; \qquad \alpha_2 P \rightarrow X.$$

Here S is the limiting substrate (which is usually a hydrocarbon); P is the product and X is the biomass (dry weight); α and β are stoichiometric constants.

The first case has been the subject of recent theoretical and experimental research. As the concentration of P increases, the specific growth rate μ decreases which indicates that μ is a function of two variables: $\mu = \mu\ (S,\ P)$. In the extended MONOD model the dependence μ of P is represented by a hyperbolic or sigmoidal curve and gives a good description of many cases of product inhibition (JERUSALIMSKY and NERONOVA 1965, KNORRE 1971, BERGTER 1972, ROMANOVSKY *et al.* 1974). Damped and undamped oscillations may be generated in the respective continuous cultures (KNORRE 1966, GÜNTHER *et al.* 1969, BERGTER 1972, 1973). If more than one inhibiting product with interrelated synthesis is considered, this results in substantially more complex kinetic models (MÜLLER *et al.* 1974).

The second case is experimentally exemplified by the formation of ethanol from glucose by *Saccharomyces cerevisiae* under aerobic conditions. Ethanol is formed in the first exponential growth phase which is followed, after a lag phase, by the second exponential phase with the ethanol as substrate but with a lower growth rate (VON MEYENBURG 1969). For an accurate description of the chemostat behaviour of the culture it was necessary to incorporate in the two-phase growth model appropriate expressions for product inhibition (BERGTER and KNORRE 1972).

The third case assumes the product to be a metabolic intermediate. With an excessive carbon source, organic acids may typically be excreted into the medium. Growth inhibition may not be necessarily due to a lower pH of the medium but may be observed at a constant pH where the intermediate is the effector of cellular regulation. Specific inhibition of this kind can be demonstrated by the simultaneous consumption of glucose and pyruvate by *Escherichia coli* (Hegewald and Knorre 1978). The present study is concerned with the structural model of the third equation $S \rightarrow P \rightarrow X$. The product P may act as an independent substrate in which case our scheme is reduced to the well-known Monod model.

We shall demonstrate that, unlike the Monod model which describes three growth phases only, viz. exponential, declining and stationary, our elementary structural model describes many observable growth processes. Here we shall discuss three variant forms of the model: A — the intermediate P has no regulatory function; B — P serves as inhibitor of its own synthesis, a process known as feedback inhibition; and C — P inhibits the formation of X. This last case corresponds to substrate inhibition, or feedforward inhibition.

When interpreted in the sense of the bottle-neck principle by Jerusalimsky (1967) our models mean that the overall population growth kinetics will be defined by either the first or the second step of our scheme. With the problem thus stated it is possible to pass over from one cell "bottle-neck" to another during culture growth, *i.e.*, it is possible to describe unbalanced population growth.

These three models enlarged for continuous culture will be discussed in following papers (Knorre *et al.* in preparation, Guthke *et al.* in preparation).

Model

In the structural model $S \rightarrow P \rightarrow X$, S is the primary substrate; it forms the intermediate P which serves as an independent substrate to synthesize the biomass X. We assume that S, P, and X correspond to the concentration of these substances evenly distributed over the culture volume. With $S(0) = 0$ conversion to the Monod model is possible. Let us further assume that the two reactions are catalysed by enzymes whose quantity is proportional to the biomass X.

We shall consider the changes in time of each substance pool. The concentration of P increases at the rate of its production in the first reaction; it decreases at the rate of its consumption by the second reaction.

The basic equations of the mathematical model can be written now:

$$\dot{S} = -\frac{1}{Y_1} K_p\,(...)\,X \tag{1}$$

$$\dot{P} = K_p(...)\,X - \frac{1}{Y_2}\mu(...)\,X \tag{2}$$

$$\dot{X} = \mu(...)\,X \tag{3}$$

where K_p (...) is the product formation rate, and μ (...) is the growth rate. The relations of K_p and μ to S and P will depend on our assumptions as to the reaction regulatory mechanisms. Y_1 and Y_2 are stoichiometric constants which indicate the weight in grammes of the respective yields per gramme of the reactant.

The material balance equation of the model is:

$$\frac{\mathrm{d}}{\mathrm{d}t}(Y_1 Y_2 S + Y_2 P + X) = 0$$

Hence

$$Y_1 Y_2 S(t) + Y_2 P(t) + X(t) = Y_1 Y_2 S(0) + Y_2 P(0) + X(0) = \text{const.}$$

For batch culture, there is a single stable stationary state

$$(\bar{S}, \bar{P}, \bar{X}) : \bar{S} = 0; \ \bar{P} = 0; \ \bar{X} = X(0) + Y_2 P(0) + Y_1 Y_2 S(0)$$

It can be seen from the right parts of the system of equations (1) and (3) that $\dot{S}$ is always negative and $\dot{X}$ is always positive. This means that $S(t)$ monotonically decreases from $S(0)$ to 0 and $X(t)$ monotonically increases from $X(0)$ to $\bar{X}$. $P(t)$ depends in its behaviour on the value of $P(0)$: it may either monotonically decrease to zero or have one maximum.

It does not seem possible to obtain analytically explicit curves of growth, $X(t)$, substrate consumption, $S(t)$, and intermediate variation, $P(t)$. We studied the behaviour of the system using the technique of qualitative investigation of the phase trajectories (ROMANOVSKY *et al.* 1974).

On the phase plane, P, S each phase trajectory $P(S)$ represents the change in the substrate and the intermediate with culture growth, and the beginnings of the phase trajectories represent the initial conditions of inoculation, $P(0)$ and $S(0)$. The set of phase trajectories is defined by the equation

$$\frac{\mathrm{d}P}{\mathrm{d}S} = -\frac{Y_1 Y_2 K_p(\ldots) - Y_1 \mu(\ldots)}{Y_2 K_p(\ldots)} \tag{4}$$

Phase trajectories have singular points where the numerator of Eqn. (4) is zero. At those points $\dot{P} = 0$, *i.e.*, $P(t)$ also has a maximum or a bending point. The locus of the phase trajectory singular points will be referred to as the separating curve. This divides the phase plane into domains where $P > 0$ ($P(t)$ increases) and $P < 0$ ($P(t)$ decreases).

The qualitative behaviour of $X(t)$ also can be investigated on this phase plane since the specific growth rate influencing the behaviour of the growth is uniquely connected with P.

In addition to the qualitative analytical study by the phase trajectory the mathematical model (1) — (3) was simulated on an analog computer.

Results

Model A

In this version of our structural model we assume that the intermediate P does not have any specific regulatory effect on cell "bottle-necks". We take that the substrate dependence of reaction rate obeys the MICHAELIS-MENTEN equation (DIXON and WEBB 1964). Then we substitute the following expressions for reaction rate in the system of equations (1) — (3):

$$K_p(\ldots) = K_{p\max} \frac{S}{K_1 + S} \tag{5}$$

$$\mu(\ldots) = \mu_{\max} \frac{P}{K_2 + P} \tag{6}$$

By substituting (5) and (6) in (4) and equating (4) to zero we obtain an expression for the separating curve:

$$P(S) = \frac{Y_2 K_{p\max} K_2 S}{K_1 \mu_{\max} + S(\mu_{\max} - Y_2 K_{p\max})} \tag{7}$$

This is a hyperbolic curve which may have a vertical or a horizontal asymptote depending on the relationship between parameters (Fig. 1a, b, d). For $\mu_{\max} \to Y_2 K_{p\max}$ the asymptotes extend to infinity and the separating curve approaches a line (Fig. 1c).

The behaviour of the phase trajectories, and consequently the kinetics of consumption of S and variation of $P(t)$, will be determined by the initial conditions and the combination of model parameters. For all initial conditions where the tangent to

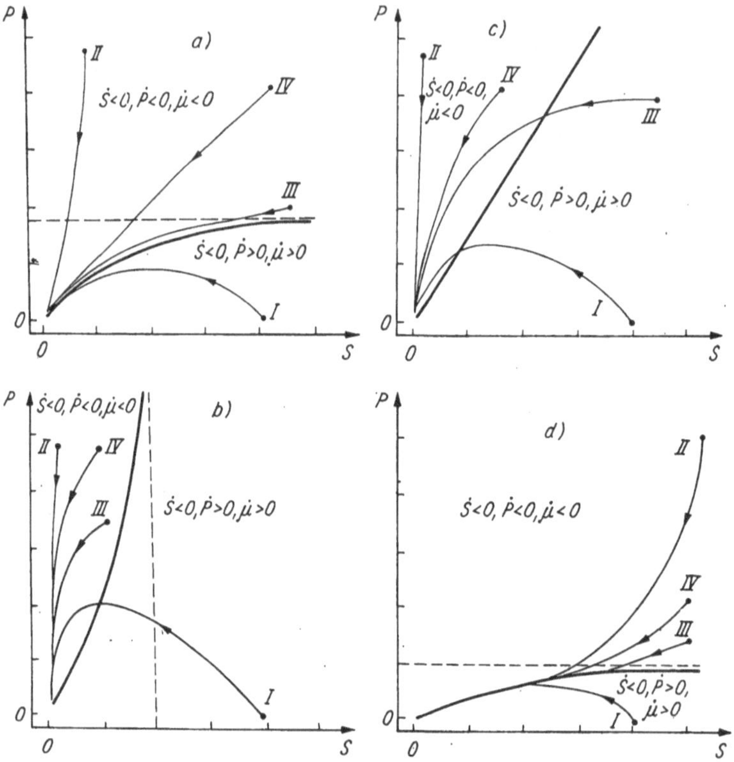

Fig. 1. Model A phase patterns.
(a) for $\mu_{\max} - Y_2 K_{p\max} > 0$,
(b) for $\mu_{\max} - Y_2 K_{p\max} < 0$,
(c) for $\mu_{\max} - Y_2 K_{p\max} = 0$,
(d) for $\mu_{\max} - Y_2 K_{p\max} > 0$, but other than in (a)
Notation: solid unbroken lines signify separating curves, fine unbroken lines signify phase trajectories, arrows indicate phase trajectory directions; dashed lines are separating curve asymptotes. For explanations of phase trajectories I, II, III, and IV see the text

the phase trajectory at an initial instant is inclined, *i.e.*, the expression

$$\operatorname{tg} \alpha = \frac{\mu_{\max}}{K_{p\max}} \frac{Y_1}{Y_2} \frac{P(0)}{K_2 + P(0)} \frac{K_1 + S(0)}{S(0)} - Y_1 < 0$$

is satisfied, S will be monotonically exhausted and $P(t)$ will have a maximum (type I of phase trajectories, in Fig. 1, for behaviour kinetics, see Fig. 2a).

If at an initial instant $1 < \operatorname{tg} \alpha < \infty$, then P monotonically decreases and exhausts itself quicker than S (type II of phase trajectories, in Fig. 1; for behaviour kinetics, see Fig 2b). If $0 < \operatorname{tg} \alpha < 1$, then S is consumed quicker than P (type III of phase trajectories in Fig 1). It shall be noted that for tg $\alpha \approx 1$ there is a simultaneous consumption of the substrate and the intermediate (type IV) phase trajectories in Fig. 1; (for behaviour kinetics, see Fig 2d). If tg $\alpha \approx 0$ then the intermediate is almost not consumed at all for a long time. This phase trajectory is the separating curve itself (Fig. 1a and d) as well as the set of phase trajectories in Fig. 1 when $P(0)$ is large enough (for behaviour kinetics, see Fig. 2c).

With a change in system parameters there is a shift in the type IV trajectory tg $\alpha = 1$ which separates the initial condition domains having different consumption kinetics for S and P (Fig. 1a and d). The phase trajectories quickly reach the separating curve and move along it (Fig. 1d, type II of phase trajectory). $\dot{P} = 0$ on the separating curve, consequently, a quasi-stationary area will be observed in the curve of consumption of P (Fig. 2b).

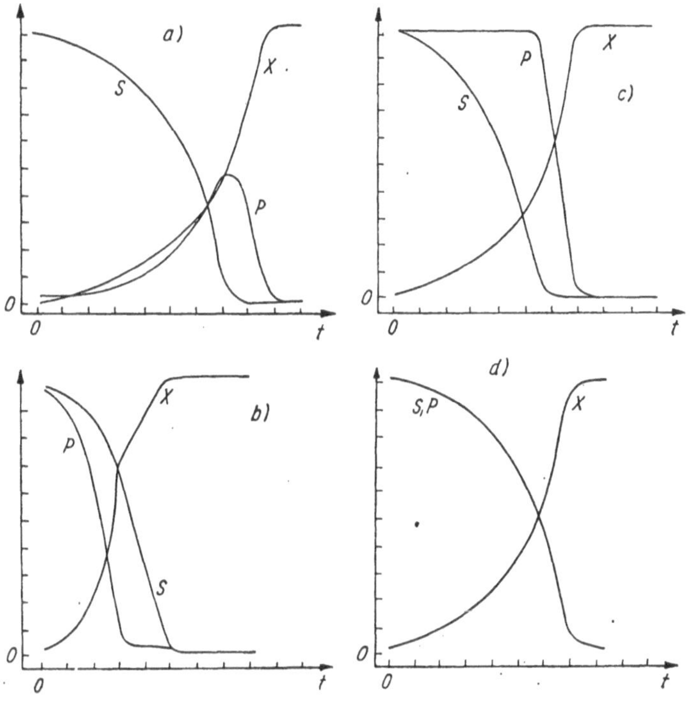

Fig. 2. Model A consumption kinetics of S, variation of $P(t)$, and growth curves.
(a) sigmoidal growth curve (type I of phase trajectories of Fig. 1).
(b) two-phase growth curve, P exhausts itself faster than S (type II of phase trajectories of Fig. 1).
(c)—(d) curves with a long exponential growth phase but different consumption of P and S (type III and IV phase trajectories of Fig. 1)

In order to obtain a qualitative pattern of behaviour of $X(t)$ let us examine change in the behaviour of $\mu = \dot{X}/X$. It can be demonstrated that

$$\dot{\mu} = \frac{\mu_{\max} K_2}{(K_2 + P)^2} \cdot \dot{P} \tag{8}$$

It follows that $\dot{\mu} \approx 0$ in two domains: at large $P (P \gg K_2)$ and at $\dot{P} = 0$ (on the separating curve). It also follows that $\dot{\mu}$ has the same sign as $\dot{P}$. We shall note that for $\dot{\mu} = 0$ ($\mu = $ const.) there is an exponential phase in the growth curve.

Fig. 1 shows that model A leads to three types of growth curves: a sigmoidal curve with an acceleration, an exponential and a growth retardation phase (type I of phase trajectories and growth curve in Fig. 2a), a curve with a long exponential growth phase (Fig. 2c, d); and a curve with two alternating exponential growth phases (type II of phase trajectory in Fig. 1d), the former induced by the initial conditions ($P \gg K_2$) and the latter arising during the approach to the sepraráting curve (the growth curve in Fig. 2b)

It should be mentioned that the curve with a long phase of exponential growth may correspond to two distinct phase trajectories: firstly, for the initial conditions $P \gg K_2$ in the type II phase trajectories of Fig. 1a, and secondly, for the initial conditions close to the separating curve. The observed maximal specific growth rate of the culture will differ (Fig. 3).

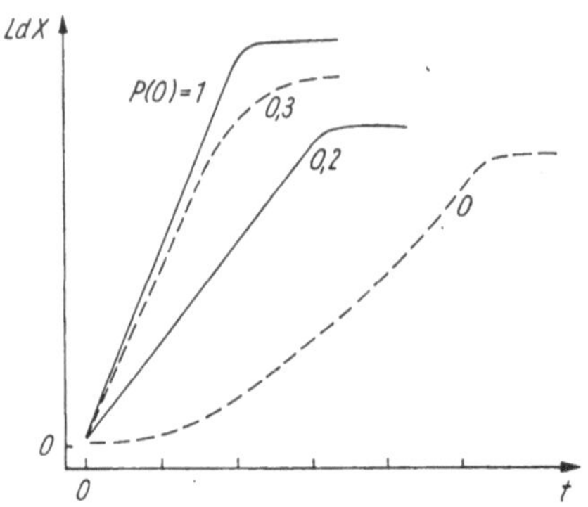

Fig. 3. Model A growth curves (ld X) for different $P(0)$. Solid curves signify exponential growth phases with distinct maximal growth rates

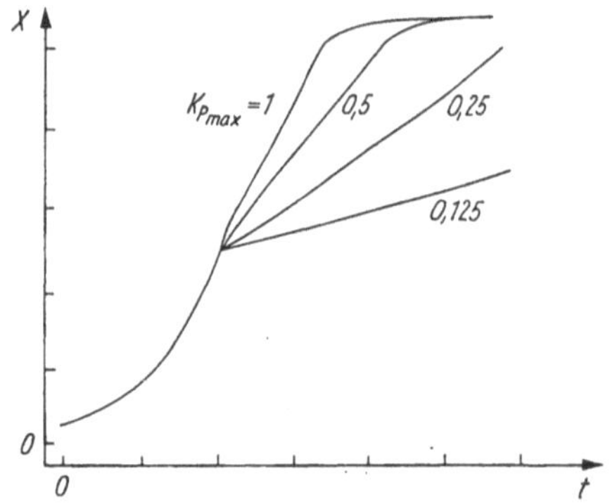

Fig 4. Model B growth curves for different maximal product formation rates

Model B

If the intermediate inhibits the reaction rate of substrate consumption then, in accordance with enzyme kinetics laws,

$$K_p(...) = K_{p\max} \frac{S}{K_1 + S} \frac{K_3}{K_3 + P} \tag{9}$$

and $\mu(...)$ corresponds to Eqn. (6). By substituting (9) and (6) in Eqn. (4) and equating it to zero we obtain an expression for the model B separating curve. It is more complex than for model A. It can be shown, however, that the model B separating curve is hyperbolic with a horizontal asymptote, *i.e.*, that it is of the type of the Fig. 1a separating curve. With certain combinations of parameters in the domain of low S and P an inflection point may appear on the separating curve.

This will not affect the qualitative pattern of the phase trajectories: the pattern of consumption of S, the growth curves and the behaviour of $P(t)$, in model B will correspond to those of model A for the case in Fig. 1a.

Using an analog computer we were able to trace the effect of variation of the parameter $K_{p\max}$ on the form of the growth curve in the presence of two successive exponential growth phases. It was found that the maximal rate of product formation affected the second phase only: as $K_{p\max}$ decreased, so did the growth rate in the second exponential phase (Fig. 4). This means that the second reaction acts as metabolism "bottle-neck" during the first hours of growth, after which the growth is determined by the first reaction.

Model C

This version of the structural model is based on the assumption that there is substrate inhibition by the intermediate of the reaction leading to X. Then the expression for the intermediate formation rate will be as (5) and

$$\mu(\ldots) = \mu_{\max} \frac{P}{K_2 + P} \frac{K_3}{K_3 + P} \cdot \tag{10}$$

A general expression for the separating curve can be obtained for this model as well. Two types of separating curves are possible depending on the parameters: with two horizontal asymptotes and a curve with a maximum at the point

$$P = \sqrt{K_2 \cdot K_3} \qquad \text{(Fig. 5).}$$

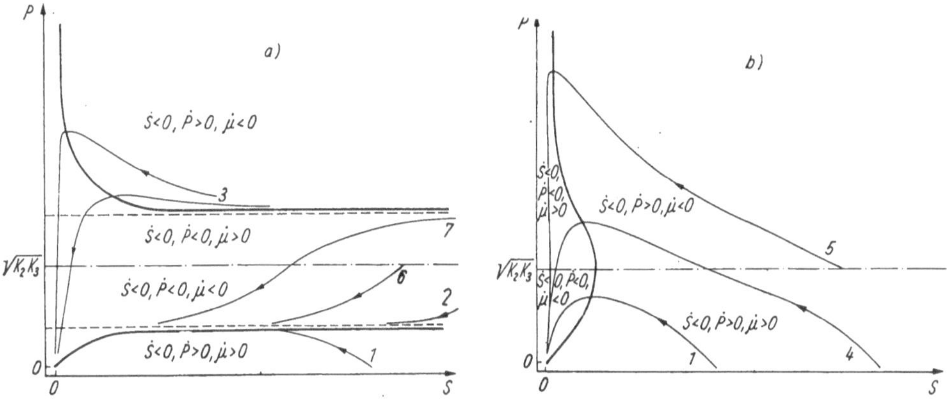

Fig. 5. Model C phase patterns.

a) for $\frac{\mu_{\max}}{Y_2 K_{p\max}} > \left(1 + \sqrt{\frac{K_2}{K_3}}\right)^2$

(b) other than (a).

Notation: The dot-and-dash line separates the change-of-sign domains of μ. Others as in Fig. 1

The increase and decrease domains of $P(t)$ ($\dot{P} \lessgtr 0$) are marked in Fig. 5. On the separating curve $\dot{P} = 0$.

Cases of distinct consumption kinetics of S and P were investigated in quite the same way as in model A: clearly, the tangent to the phase trajectory at an initial instant defines the behaviour of $P(t)$ as well as supplies an answer to the question whether S or P will be the first to be exhausted.

It can be seen in Fig. 5b that for all the initial conditions in the phase plane to the right of the separating curve, $P(t)$ will have its maximum; for the initial conditions to the left of the separating curve, $P(t)$ will monotonically decrease. Fig. 5a also clearly demonstrates the pattern of variation of $P(t)$: for the initial conditions under the lower branch and above the upper branch of the separating curve the phase trajectories and $P(t)$ have a maximum, and between the separating curve branches the phase trajectories and $P(t)$ monotonically decrease. Of particular interest is the domain

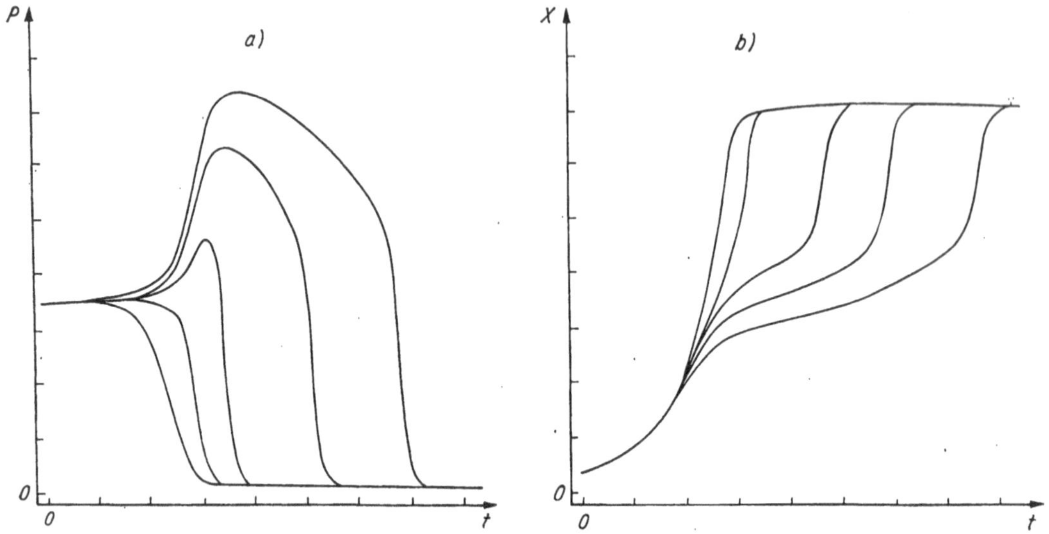

Fig. 6. Behavioural variation in $P(t)$ and growth curves with slight variations in the initial conditions $P(0)$ in domains close to the upper branch of the model C separating curve.
(a) behaviour of $P(t)$, (b) growth curves $X(t)$

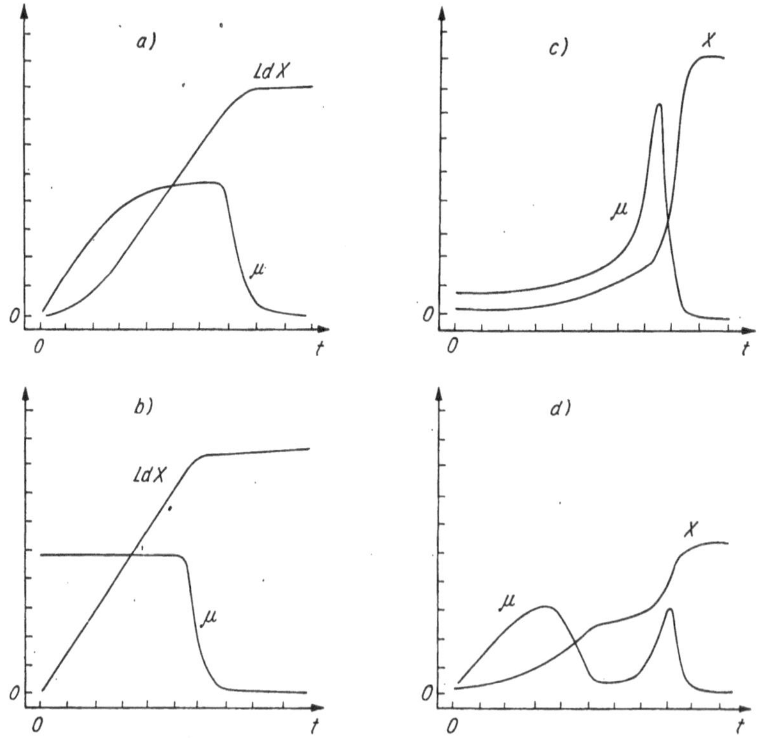

Fig. 7. Model C growth curves and specific growth rate behaviour.
(a) sigmoidal growth curve (phase trajectory 1 of Fig. 5.), (b) growth curve with a long exponential phase (phase trajectory 2 of Fig. 5), (c) growth curve with a long lag phase (phase trajectory 3 of Fig. 5), (d) growth curve with diauxy (phase trajectory 4 of Fig. 5)

of initial conditions located near the upper branch of the separating curve: slight variation in the starting conditions may cause significant changes in the behaviour of $P(t)$ (Fig. 6a).

We shall examine the behaviour of $\mu(t)$ in order to have an idea of the growth curves that can be obtained with this model. It can be demonstrated that

$$\dot{\mu} = \frac{\mu_{\max} K_3}{(K_2 + P_2)^2 (K_3 + P)^2} \dot{P}(K_2 K_3 - P^2) \tag{11}$$

The exponential growth phase will thus be observed in three domains of the phase plane P, S: where $P \gg \sqrt{K_2 K_3}$; where $\dot{P} = 0$ (near the separating curve), and where $P = \sqrt{K_2 \cdot K_3}$. On the two last lines $\dot{\mu}$ will change sign. Different sign-domains are indicated in Fig. 5.

It can be seen from the phase trajectories discussed above that model C offers different opportunities for diverse growth curves, from the already familiar model A sigmoidal curves (phase trajectory 1 in Fig. 5 and the growth curve in Fig. 7a) and curves with a long exponential phase (phase trajectory 2 in Fig. 5a, the growth curve in Fig. 7b) to curves with a long lag phase (phase trajectory 3 in Fig. 5a and the growth curve in Fig. 7c) and diauxic type curves (phase trajectory 4 in Fig. 5b and the growth curve in Fig. 7d).

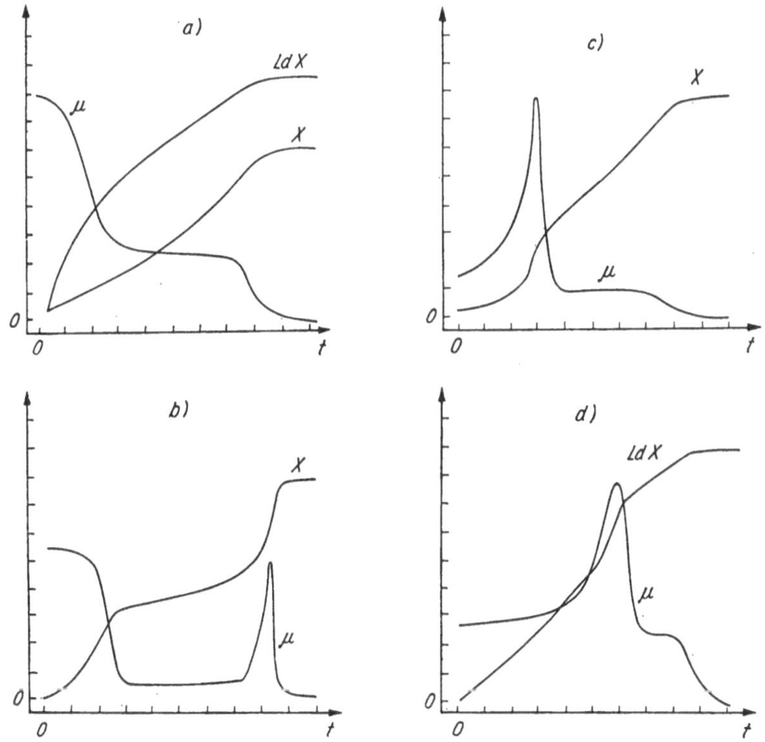

Fig. 8. Model C growth curves and growth rate behaviour versus initial conditions.
(a) corresponds to phase trajectory 6 of Fig. 5a,
(b) corresponds to phase trajectory 5 of Fig. 5b,
(c) corresponds to phase trajectory 7 of Fig. 5a,
(d) corresponds to phase trajectory 7 of Fig. 5a for parameters other than (c)

This model can explain the existence of growth curves without pronounced growth phases. This kind of dynamics is frequently observed in experiments, facing the investigator with the problem of how to compare the results with the classic growth curve. Curves of this type obtained from model C are illustrated in Fig. 8a, b, c, d: (phase trajectories 6, 5, 7 in Fig. 5, respectively).

In the initial conditions domain close to the upper separating curve, slight variation in $P(0)$ results in substantial changes in $X(t)$ (Fig. 6b).

Discussion

Experimental growth curves seldom will be of a shape corresponding to balanced growth and the theoretical MONOD equation. They are much more commonly found to deviate from it, the deviations customarily being neglected by experimentators as measuring errors or errors of experimental design. Results obtained in this study with the help of structural models can help to account for the origins of these deviations, to predict the growth curve shape and type. From the present study one is led to the conclusion that the shape of a growth curve depends on: (1) the kind of microorganism and its biochemical features, *i.e.*, the nature of interrelations between the cell "bottle-necks", and (2) the initial conditions of the inoculation, *i.e.*, culture prehistory.

Let us dwell on each factor separately. From the mathematical point of view the nature of the interrelation and microorganism type mean that a particular model must be chosen and the constants Y_i, $K_{p\max}$, $\mu_{\max}$, K_i must be quantitatively defined. This can be done after a study of the biochemical characteristics of the microorganism in question and the growth process. The primary tasks here are to interpret the variable P, the two possible cell "bottle-necks", and to develop a structural model. Examples of units which control cell enlargement generally are the catabolic and anabolic processes in heterotrophic microorganisms, the decomposition of organic substance, and the synthesis of macromolecules. In this case ATP could act as P, the substance linking and regulating the two processes.

This, of course, is just one possible interpretation; one should be guided in each particular case by the biochemical characteristics of the specific culture, and separate the system of cell biochemical reactions into units, subsystems, using previously obtained data for the relevant microorganisms.

Another possible interpretation can be to consider the first "bottle-neck" as the substrate transport process into the cell. In this case P should be interpreted as the cellular concentration of the substrate, and the relation of the first reaction rate to the extracellular substrate concentration $K_p(S)$ will vary depending on our assumptions as to transport kinetics.

Still another interpretation can be illustrated by the case of *Hydrogenomonas eutropha Z—1*, bacteria in which possible "bottle-necks" may refer to ATP production in energy exchange reactions and ATP consumption in synthesis reactions, constructive exchange reactions. It follows from the model that a lengthy quasilinear part of the growth curve, which is characteristic of hydrogen bacteria (ZAVARZIN 1972), is conditioned by a monotonous decrease in P. The data of ATP dynamics corresponded to the model conclusion as to the behaviour of P (PETROVA *et al.* 1976).

The present study also emphasizes the importance of culture prehistory in the experiment. We can obtain a variety of growth curves depending on the initial conditions $P(0)$. If P is taken to mean an extracellular substance we can set up the initial conditions ourselves thereby affecting the shape of the growth curves. If, on the other hand, P is an intracellular intermediate product, then $P(0)$ refers to the prehistory of our culture, namely, the growth phase, the number of transfers, the temperature of the inoculate, *etc.*

One example is the previously mentioned cultivation of *Escherichia coli* on glucose and pyruvate. Pyruvate is known to be an intermediate of glucose metabolism in heterotrophic bacteria. For variations in the initial conditions (starting glucose concentration S and pyruvate concentration P in the medium) different patterns of consumption of S and P and growth curves confirming the model were obtained (HEGEWALD and KNORRE 1978).

A few more conclusions from the model which may be of use in experimental work:

The model indicates that the length of the lag phase decreased with inoculum size. Experimental data (SHIDA *et al.* 1975) bear out this conclusion. The model can be used to predict the length of the lag phase for different inoculation volumes.

In work with microorganisms "non-reproducibility" of experiments in terms of model results can prove to be a consequence of a poorly chosen point of starting conditions. As has been shown for Model C, negligible deviations in $P(0)$ will lead to noticeable changes in the shape of the curve $P(t)$ and the growth curve $X(t)$ (Fig. 6). Such errors can be eliminated if the experiment's initial conditions are altered, though not in $S(0)$ but in $P(0)$.

Finally, through these same initial conditions, culture prehistory may account for the controversial data of maximum specific growth rate of identical strain cultures reported by different authors. As shown in the model (Fig. 2e), two marked growth rates in the long exponential growth phase are observed depending on $P(0)$.

On the other hand, the nature of the growth curves themselves may suggest the kind of relation between the cell bottle-necks and the direction of biochemical research. To illustrate this, diauxic growth curves (with a single substrate source) are only possible in the case of feedforward inhibition (Model C). It should be mentioned that such growth curves are observed in thermophilic microorganisms (TSAPLINA and JEGOROVA 1972).

Therefore, one is led to conclude from the reported study that the variety of growth curves and the non-reproducibility of results are due to the pattern of links between the governing cell substructures and cell prehistory. By knowing the pattern and having found the values of the model constants one is able to predict the growth dynamics and substrate consumptions in batch cultivation. And conversely, a knowledge of growth curves will help to make somed eductions concerning the relationships among individual cell units and choose a direction for further experimentation.

References

BERGTER, F., 1972. Wachstum von Mikroorganismen. Experimente und Modelle. VEB Gustav Fischer Verlag Jena.

BERGTER, F., 1973. Produktinduzierte Oszillationen der Zelldichte von *Streptococcus pyogenes* im Chemostaten. Z. Allg. Mikrobiol., **13**, 725–727

BERGTER, F. und KNORRE, W. A., 1972. Computersimulation von Wachstum und Produktbildung bei *Saccharomyces cerevisiae*. Z. Allg. Mikrobiol., **12**, 613–629.

DIXON, M. and WEBB, E. C., 1964. In: Enzymes, 2nd ed., p. 450. Longmans, Green & Co. Ltd. London.

GÜNTHER, H. H., KNORRE, W. A. and BERGTER, F., 1969. The possibility of autonomous synchronization of cell division in continuous culture of bacteria. Proc. of the 4th Symp. Continuous Cultivation of Microorganisms. Prague.

HEGEWALD, E. M. and KNORRE, W. A., 1978. Kinetics of growth and substrate consumption of *Escherichia coli* ML 30 on two carbon sources. Z. Allg. Mikrobiol., in press.

JERUSALIMSKY, N. D., 1967. Bottle-necks in metabolism and growth rate controlling factors. Third International Symp. Microbiol. Physiology and Continuous Culture.

JERUSALIMSKY, N. D. and NERONOVA, N. M., 1965. The relationship between the concentration of metabolic products and the growth rate of microorganisms (russ.). Dokl. AN SSSR, **161**, 1437.

KNORRE, W. A., 1966. Oszillationen in biologischen Systemen und ihre Beschreibung durch die „talandische" Temperatur nach GOODWIN. studia biophysica, **1**, 33–40.

Knorre, W. A., 1971. Analogcomputer in Biologie und Medizin. Einführung in die dynamische Analyse biologischer Systeme. VEB Gustav Fischer Verlag Jena.
v. Meyenburg, K.,, 1969. Katabolit-Repression und der Sprossungszyklus von *Saccharomyces cerevisiae*. Vierteljahresschrift der Naturforsch. Gesellschaft in Zürich, **114**, 113.
Müller, P. J., Bergter, F., Günther, H. H. und Knorre, W. A., 1974. pH-induzierte Wachstumsinhibition in Turbidostatenkulturen von *Escherichia coli* 2000. Z. Allg. Mikrobiol., **14**, 707—712.
Petrova, G. V., Romanova, A. K. and Malyan, A. N., 1976. ATP content in the cells of hydrogen bacterium *Hydrogenomonas eutropha* (russ.). Mikrobiologia, **45**, 178 —179.
Romanovsky, J. M., Stepanova, N. V. und Chernavsky, D. S., 1974. Kinetische Modelle in der Biophysik. VEB Gustav Fischer Verlag Jena.
Shida, T., Komagata, K. and Mitsugi, K., 1975. Reduction of lag time in bacterial growth. I. Effect of inoculum size and nutrients J. gen. appl. Microbiol., **21**, 75—86.
Tsaplina, A. and Jegorova, L. L., 1972. Some properties of growth, development and synthesis of extracellular protease by the thermophilic bacterium *Bacillus brevis* (russ.). Mikrobiologia, **41**, 798—804.
Zavarzin, G. A., 1972. Lithotrophic Microorganisms (russ.). Nauka Moscow.

Mailing addresses: Dr. T. A. Petrova
Institute of Microbiology USSR
Academy of Sciences, Moscow 117312

Dr. sc. W. A. Knorre
Zentralinstitut für Mikrobiologie und experimentelle Therapie Jena der AdW
DDR 69 Jena, Beuthenbergstraße 11

Zeitschrift für Allgemeine Mikrobiologie	17	7	1977	543–547

(Sektion Biologie — Mikrobielle Biochemie — Friedrich-Schiller-Universität Jena)

Physiologie und Biochemie der Streptomyceten

XIII. Biosynthese von Paromomycin unter Einsatz von ^{14}C-Glucose, -Glucosamin, -2-Desoxystreptamin und -Ribose durch *Streptomyces albus var. metamycinus nov. var.*

G. REUTER, H. KÖSTER und B. LIEBERMANN

(*Eingegangen am 12. 8. 1976*)

Distribution of radioactivity in paromomycin ascertained after application of ^{14}C-D-glucose, ^{14}C-D-glucosamine, ^{14}C-2-deoxystreptamine, respectively, ^{14}C-D-ribose is taken as basis for a biosynthesis scheme: While ribose bound in the antibiotic originates from glucose by oxidation and following decarboxylation, glucosamine is formed via fructose-6-phosphate. Paromose I arises from glucosamine, but not the cyclohexan derivative 2-deoxystreptamine, whose biosynthesis pathway is directly branching off glucose.

Die Biosynthese von Paromomycin kann als Modell dienen für die Bildung zahlreicher anderer Antibiotica, die Aminozucker zum Teil glycosidartig verknüpft in sich vereinigen. Der Paromomycin-Komplex besteht aus zwei Stereoisomeren — jeweils aus vier Komponenten aufgebaut — die offensichtlich in ihrer Biosynthese sehr enge Wechselwirkungen aufweisen.

Bei Applikation von D-Glucose-u-^{14}C an *Streptomyces albus var. metamycinus nov. var.* fanden wir für Paromomycin II eine stärkere radioaktive Markierung und eine andere Abhängigkeit von der Applikationszeit als für Paromomycin I, das am Paromose-Teil isomer ist (KÖSTER *et al.* 1977). Für die zwei Paromose-Isomere wurden unterschiedliche Syntheseraten angenommen, die sich nichtproportional zueinander ändern. Die Verteilung der Radioaktivität auf die Komponenten von Paromomycin I zeigte, daß die Glucose-Kette während der Biosynthese von Glucosamin, Ribose und Paromose I nicht fragmentiert wird. Dies gilt jedoch nicht für 2-Desoxystreptamin (KÖSTER *et al.* 1977).

Die zwei im Paromomycin I verbundenen Einheiten, Paromamin und Paromobiosamin sowie deren Untereinheiten, D-Glucosamin und 2-Desoxystreptamin bzw. D-Ribose und Paromose I werden unterschiedlich aus dem Nährmedium aufgenommen und metabolisiert. Glucosamin wird gut, Desoxystreptamin und Ribose bedeutend weniger und Paromose I nicht nachweisbar in Paromomycin eingebaut. Im Zusammenhang mit diesen Ergebnissen konnten wir feststellen, daß Paromamin und Paromobiosamin offensichtlich intrazellulär zunächst in ihre Untereinheiten gespalten werden (LIEBERMANN *et al.* 1977).

Durch Abbau von Paromomycin, das nach Applikation seiner ^{14}C-markierten Bausteine erhalten wurde, sollten in der vorliegenden Arbeit weitere Hinweise auf die Biosynthese von Paromomycin gewonnen werden.

Material und Methoden

Kulturführung, Zusammensetzung der Nährmedien und Isolierung des Paromomycins sowie dessen Abbau, die Gewinnung seiner radioaktiv markierten Komponenten, ihre quantitative Bestimmung und Radioaktivitätsbestimmung erfolgten nach REUTER *et al.* (1975), LIEBERMANN *et al.* (1975, 1977).

Ergebnisse und Diskussion

Nach Applikation von ^{14}C-D-Glucose, ^{14}C-D-Glucosamin, ^{14}C-2-Desoxystreptamin bzw. ^{14}C-D-Ribose wurde Paromomycin isoliert, acetyliert und in die Stereoisomeren aufgetrennt. N-Pentaacetylparomomycin I wurde hydrolysiert und die vier erhaltenen Bausteine isoliert. Die ermittelte spezifische Radioaktivität wurde den in Tabelle 1 aufgeführten Werten zugrunde gelegt.

Tabelle 1
Relative Verteilung der spezifischen Radioaktivität auf die Komponenten von Paromomycin nach Applikation verschiedener radioaktiv markierter Vorstufen

Applikation: 29 Std. nach Fermentationsbeginn; Fermentationsdauer: 96 Std.; synthetisches Nährmedium mit 3% Stärke (NM 2); die Verhältnis- bzw. Prozentzahlen ergeben sich aus den gemessenen, auf C-Atom bezogenen spezifischen Radioaktivitäten; bei Applikation von Glucose wurde die spezifische Radioaktivität von Glucosamin, sonst die der applizierten Verbindung mit 1,00 angesetzt; $GlcNH_2$ = Glucosamin; DOS = 2-Desoxystreptamin

applizierte ^{14}C-Verbindungen	isolierte Verbindungen			
	$GlcNH_2$	DOS	Ribose	Paromose
D-Glucose	1,00 25,3%	0,91 23,0%	1,12 28,3%	0,90 23,4%
D-$GlcNH_2$ · HCl	1,00 56,1%	0,004 0,23%	0,006 0,37%	0,77 43,3%
2-DOS · 2 HCl	0,03 2,6%	1,00 87,7%	0,11 9,7%	— —
D-Ribose	0,34 17,2%	0,29 14,6%	1,00 50,3%	0,35 17,7%

Während bei Applikation von Glucose eine annähernd uniforme Verteilung auf alle 4 Komponenten erfolgt, treten bei Glucosamin-Fütterung bemerkenswerte Unterschiede auf. Die Radioaktivität ist hauptsächlich auf Glucosamin und Paromose verteilt. Ribose, jedoch auch 2-Desoxystreptamin zeigen nur geringfügige Markierung. Die Radioaktivität des applizierten 2-Desoxystreptamins befindet sich zum größten Teil im 2-Desoxystreptamin des isolierten Paromomycins, während bei Ribose-Fütterung die Hälfte der Radioaktivität in der aus Paromomycin isolierten Ribose aufgefunden wird. Der Rest verteilt sich annähernd gleichmäßig auf die übrigen Verbindungen.

Die uniforme Verteilung der Radioaktivität nach Fütterung von D-Glucose-u-^{14}C war zu erwarten: Stärke war die einzige C-Quelle im Nährmedium. Eine Verbindung zwischen Ribose und Glucose bzw. Fructose-6-phosphat über den Hexosemonophosphat-Weg könnte das ähnliche Verteilungsmuster bei Applikation von Ribose erklären. Für die Bildung der Ribose im Neomycin werden zwei Wege vorgeschlagen: a) Oxydation der Glucose über den Hexosemonophosphat-Weg unter Decarboxylierung am C 1 und b) Glucuronat-Weg unter Decarboxylierung am C 6. Experimente mit Glucose-1-^{14}C-, Glucosamin-1-^{13}C und Glucose-6-^{13}C sprechen für beide Möglichkeiten (Schimbor 1968, Rinehart *et al.* 1974).

Glucosamin kann offensichtlich weder zur Biosynthese von Ribose noch von 2-Desoxystreptamin verwendet werden. Für die Biosynthese von Glucosamin-6-phosphat aus Fructose-6-phosphat und Glutamin scheint demnach im Zusammenhang mit der Paromomycinbiosynthese kein reversibler Reaktionsweg zu existieren. In Paromose jedoch erfolgt nach Glucosamin-Applikation ausgesprochen guter Einbau.

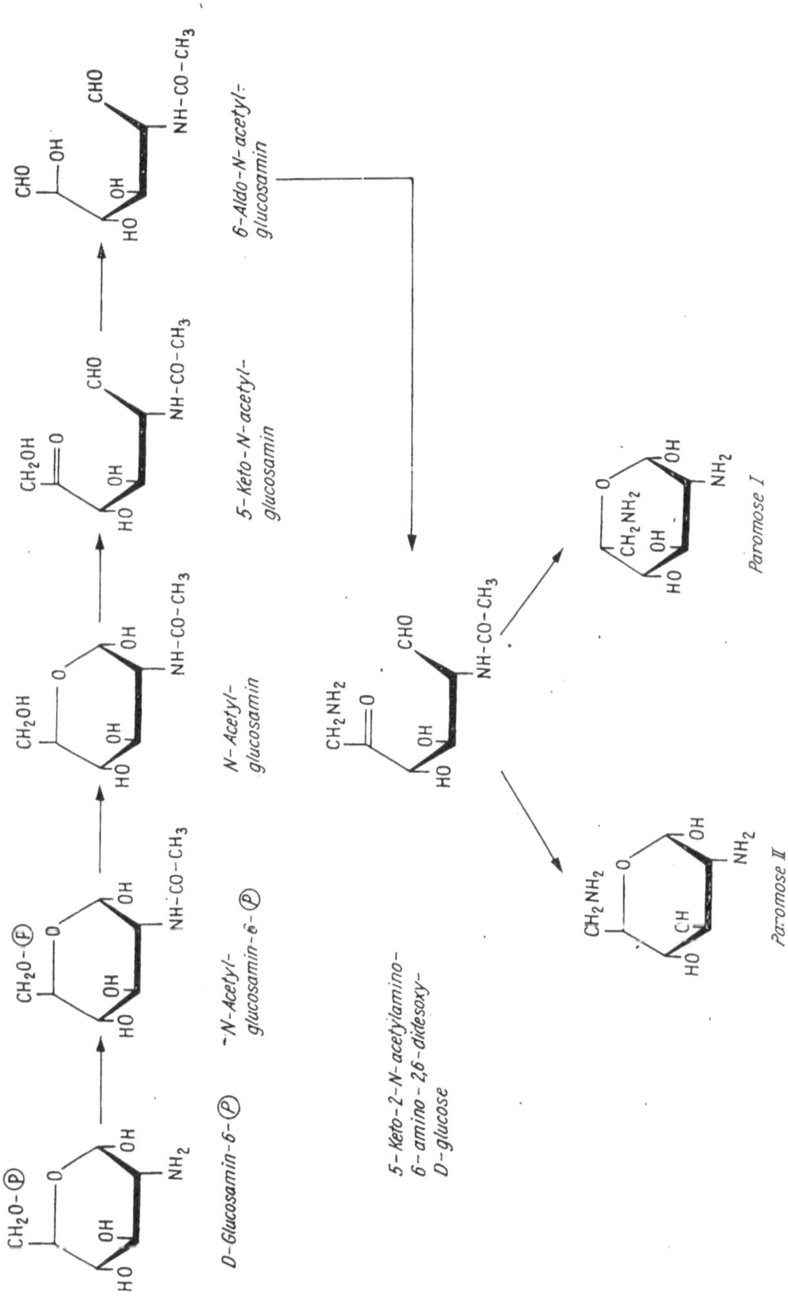

Schema 1. Paromose-Biosynthese aus Glucosamin

In Abwandlung der Vorstellungen über die Biosynthese der Neomycin-Bausteine (FOGHT 1963, RINEHART u. SCHIMBOR 1967) nehmen wir die in Schema 1 dargestellte Reaktionsfolge über Glucosamin und eine 2,6-Diamino-5-oxohexose an. Die hypothetischen Verbindungen zwischen Glucosamin-6-phosphat und den Diaminohexosen sind in Schema 2 als × bezeichnet.

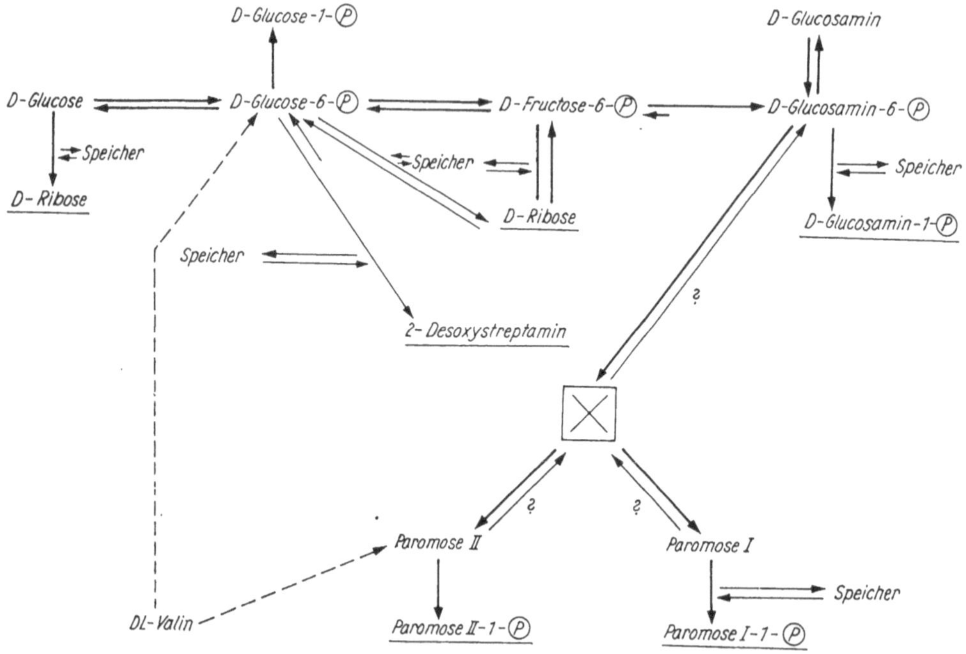

Schema 2. Biosynthese der Paromomycin-Bausteine

Im Gegensatz zur Vorstellung, daß bei der Neomycin-Biosynthese Glucosamin auch für Desoxystreptamin als Vorstufe fungiert, sprechen unsere Einbauversuche in das Paromomycin nicht für diesen Weg. Es kann diskutiert werden, daß Desoxystreptamin aus dem Glucose-Pool unter Fragmentierung des Kohlenstoffskeletts über eine Reihe unbekannter Schritte entsteht. Bereits vorangegangene Markierungsversuche (KÖSTER *et al.* 1977) weisen in die gleiche Richtung: In einem komplexen organischen Fermentationsmedium wird das Desoxystreptamin des gebildeten Paromomycins durch Glucose-u-^{14}C weniger stark markiert als die anderen Bausteine.

Als weitere Möglichkeit der Desoxystreptamin-Biosynthese ist der von RINEHART *et al.* (1974) vorgeschlagene Weg zu sehen, der über eine Desoxyinosose und deren Aminierung, anschließende Oxydation und Aminierung in β-Stellung verlaufen könnte.

Die unterschiedliche spezifische Radioaktivität der einzelnen Stereoisomere bei Fütterung von D-Glucose-u-^{14}C führte zu der Vorstellung von Speichern, die im Nebenanschluß der Biosynthesewege liegen (KÖSTER *et al.* 1977). Diese Versuche ließen vermuten, daß mit Ausnahme von Paromose II alle Bausteine des Paromomycins nicht nur ausschließlich auf direktem Wege für die Biosynthese des Paromomycins verwendet werden (s. Schema 2), sondern in einem gewissen Umfang vorübergehend in bestimmte höhermolekulare Verbindungen, z. B. in solche der Zellwand eingehen.

Unter der Annahme, daß für Paromose II diese Möglichkeit nicht besteht, ist erklärlich, daß Paromomycin II eine höhere spezifische Radioaktivität zeigt als das entsprechende Stereoisomere I.

Die bei KÖSTER *et al.* (1977) beschriebenen Veränderungen der Radioaktivitätsverteilung und der Mengenverhältnisse der Stereoisomere im Paromomycinkomplex bei Biosynthese in einem Valin-Nährmedium lassen die Annahme zu, daß Valin bzw. dessen Abbauprodukte unmittelbar vor der Bildung von Paromose II in die Synthesekette eingeschleust werden und damit eine ähnliche Wirkung wie ein Speicher bei der Paromose I-Bildung haben.

Es ist zu erwarten, daß die Verknüpfung der Bausteine des Paromomycins analog der Oligosaccharid-Biosynthese über 1-Phosphoderivate und entsprechende Nucleotide erfolgt. So zeigten PAPE u. BRILLINGER (1973) die Bildung von Thymidin-diphosphat-mycarose durch ein zellfreies System aus *Streptomyces rimosus*. ORTMANN *et al.* (1974) fanden eine NADPH-abhängige Bildung von Thymidin-diphosphodihydrostreptose in einem zellfreien System von *S. griseus*.

Für zuverlässige technische Assistenz danken wir Frau B. HERRMANN und Frau K. SCHLAG.

Literatur

FOGHT, J., 1963. Biosynthetic studies on the neomycins. Ph. D. Thesis, Univ. of Illinois.

KÖSTER, H., LIEBERMANN, B. und REUTER, G., 1977. Physiologie und Biochemie der Streptomyceten XI. Unterschiedlicher Einbau von D-Glucose-u-^{14}C in die Paromomcyin-Isomere und in die Bausteine von Paromomycin I. Z. Allg. Mikrobiol., **17**, 433—436.

LIEBERMANN, B., KÖSTER, H. und REUTER, G. 1975. Physiologie und Biochemie der Streptomyceten. II. Chemischer Abbau und Isolierung der Abbauprodukte von Paromomycin. Pharmazie, **30**, 797—798.

LIEBERMANN, B., KÖSTER, H. und REUTER, G., 1977. Physiologie und Biochemie der Streptomyceten. 12. Unterschiedliche Aufnahme und Verwertung radioaktiv markierter Bausteine von Paromomycin durch *Streptomyces albus var. metamycinus novus var.* Pharmazie, im Druck.

ORTMANN, R., MATERN, U., GRISEBACH, H., STADLER, P., SINNEWELL, V. and PAULSEN, H., 1974. NADPH-dependent formation of thymidine diphosphodihydrostreptose from thymidine diphospho-D-glucose in a cell-free system from *Streptomyces griseus* and its correlation with streptomycin biosynthesis. Europ. J. Biochem., **43**, 265—271.

PAPE, H. und BRILLINGER G. U., 1973. Stoffwechselprodukte von Mikroorganismen. 113. Mitt. Biosynthese von Thymidindiphospho-mycarose durch ein zellfreies System aus *Streptomyces rimosus*. Arch. Mikrobiol., **88**, 25—35.

REUTER, G., LIEBERMANN, B. und Köster, H. 1975. Physiologie und Biochemie der Streptomyceten. 1. Isolierung, Identifizierung und Untersuchungen zur mikrobiellen Synthese der Stereoisomere im Paromomycin-Komplex. Pharmazie, **30**, 733—736.

RINEHART JR., K. L. and SCHIMBOR, R. F., 1967. Neomycins. In: Antibiotics, Vol. II (D. GOTTLIEB and P. D. SHAW, Editors). Springer-Verlag Berlin—Heidelberg—New York.

RINEHART Jr., K. L., MALIK, J. M., NYSTROM, R. S., STROSHANE, R. M., TRUITT, S. T., TANIGUCHI, M., ROLLS, J. P., HAAK, W. J. and RUFF, B. A., 1974. Biosynthetic incorporation of 1-^{13}C-glucosamine and 6-^{13}C-glucose into neomycin B. J. Amer. chem. Soc., **96**, 2263.

SCHIMBOR, R. F., 1966. The microbiological incorporation of labelled intermediates into the nemoycin antibiotics. Ph. D. Thesis, Univ. of Illinois.

Anschrift: Prof. Dr. G. REUTER
Sektion Biologie, AG Mikrobielle Biochemie
DDR 69 Jena, Wöllnitzer Str. 7

Zeitschrift für Allgemeine Mikrobiologie	17	7	1977	549–558

(Sektion Biologie, WB Allgemeine Mikrobiologie
der Ernst-Moritz-Arndt-Universität Greifswald)

DNS-Stoffwechsel einer temperatursensitiven filamentösen Mutante von *Bacillus subtilis* SB 19

D. SCHRÖDER, HILTRAUT MACH und F. MACH

(Eingegangen am 16. 8. 1976)

Die temperatursensitive filamentöse Mutante *ts* 33-6 von *B. subtilis* SB 19 zeichnet sich durch unterschiedliches Teilungsvermögen bei restriktiver und permissiver Temperatur aus (30 °C Kurzstäbchen, 46 °C Filamente). Sie verfügt unter restriktiven Verhältnissen (46 °C) über einen funktionsfähigen Replikationsmechanismus. Die während der Sporenkeimung bei 45 °C begonnene DNS-Replikation kann auch unter restriktiven Temperaturbedingungen beendet werden.

Kurzzeitige Beeinflussung gekeimter Sporen durch Inkubation bei 48 °C und anschließender shift down zu 46 °C verändern den DNS-Syntheseablauf, verhindern jedoch die Filamentbildung nicht.

Teilungsinduktion durch Temperaturwechsel von 46 °C zu 30 °C ist mit gesteigerter DNS-Synthese gekoppelt. Gegenüber einer bei 46 °C verbleibenden Kontrolle verstärkt sich die DNS-Syntheseaktivität.

Die DNS-Synthese nach Temperaturerniedrigung läuft unabhängig von der Proteinsynthesehemmung durch Chloramphenicol ab. Auch bei blockierter RNS-Synthese in aminosäurefreiem Medium ist die DNS-Replikation möglich. Lediglich Mitomycin C in geeigneten Konzentrationen hemmt bzw. beeinflußt die DNS-Replikation nach shift down.

Der Lebenszyklus von Mikroorganismen bedingt das exakt aufeinander abgestimmte Wechselspiel von Protein-, RNS-, und DNS-Synthese. Die Makromolekülsynthesen bilden die Grundlage für das Wachstum und damit verbunden auch für die Zellteilung.

Mikroorganismen mit gestörtem Teilungsmodus bieten eine Möglichkeit, das komplexe Geschehen der bakteriellen Zellteilung näher zu untersuchen. Die von verschiedenen Autoren (MENDELSON u. GROSS 1967, BAZILL u. RETIEFF 1969, MACH u. ENGELBRECHT 1970, UPCROFT *et al.* 1975, HANNA u. CARL 1975) beschriebenen temperatursensitiven Mutanten sind für derartige Untersuchungen besonders geeignet. Es handelt sich dabei um Organismen, die durch mutagene Behandlung genotypisch so verändert sind, daß sie in einem bestimmten, engbegrenzten Temperaturbereich nicht mehr in der Lage sind, sich zu teilen. Unabhängig davon laufen die Proteinbiosynthese und die RNS-Synthese mehr oder weniger normal ab. Oft sind jedoch Veränderungen unterschiedlicher Art im Bereich des DNS-Metabolismus festgestellt worden.

Die hier beschriebenen Untersuchungen wurden an einer von MACH u. ENGELBRECHT (1970) isolierten temperatursensitiven, filamentösen Mutante von *Bacillus subtilis SB* 19 durchgeführt. Im Rahmen der allgemeinen molekularbiologischen Charakterisierung dieser Mutante bestand das Ziel darin, den DNS-Metabolismus unter verschiedenen Wachstumsbedingungen zu untersuchen.

Material und Methoden

Als Untersuchungsobjekt diente die aus dem Wildstamm von *Bacillus subtilis* SB 19 nach mutagener Behandlung mit Hydroxylamin von MACH u. ENGELBRECHT (1970) gewonnene temperatursensitive, filamentöse Mutante *ts* 33-6. Dieser Organismus ist außerdem durch eine vierfache

Auxotrophie (val⁻, leu⁻, Ileu⁻, $GluNH_2^-$) charakterisiert. Bei der Durchführung aller Experimente mit der Mutante *ts* 33-6 waren Sporen das Ausgangsmaterial. Die Keimung der Sporen im modifizierten Medium nach TORRIANI u. LEVINTHAL (1967), MACH u. ENGELBRECHT (1970) sowie die Induktion der Filamentbildung im angereicherten Minimalmedium nach COPELAND u. MARMUR (1968) wurden, wie bereits SÜSS *et al.* (1972) beschrieben, durchgeführt.

Die Komponente I des angereicherten Minimalmediums nach COPELAND u. MARMUR (1968) mit der Zusammensetzung: 0,6% KH_2PO_4, 1,4% K_2HPO_4, 0,1% Natriumcitrat × 2 H_2O, 0,2% $(NH_4)_2SO_4$, 0,01% $MgSO_4$ und einem Zusatz von 0,5% Glucose diente bei speziellen Untersuchungen als aminosäurefreies Minimalmedium.

Die Kinetik der DNS-Synthese wurde durch die Inkorporation von ^3H-Thymidin, die der RNS-Synthese durch die Inkorporation von ^3H-Uracil bestimmt.

Im Verlaufe der Kultivierung des Organismus in Gegenwart des radioaktiv markierten Prekursors erfolgt zu bestimmten Zeiten die Entnahme von 1 ml Probe aus dem Versuchsansatz. Die Probe wurde mit dem gleichen Volumen kalter 10%iger Trichloressigsäure gemischt. Nach 30-minütigem Stehen bei +4 °C konnte die Trichloressigsäure-unlösliche Komponente durch Membranfiltration gesammelt werden.

Die mit aqua dest. gewaschenen und anschließend getrockneten Membranfilter wurden in kaliumfreie Glasküvetten gelegt und mit einer Scintillatorlösung bestehend aus 4,0 g PPO und 0,1 bis 0,25 g Dimethyl-POPOP, gelöst in 1000 ml Toluol, überschichtet.

Die Messung der Radioaktivität konnte mit Hilfe eines Flüssigkeitsscintillations-Spektrometers (TRICARB-Liquid-Scintillationsspectrometer Modell 3320, PACKARD Instruments Company Chicago, USA) vorgenommen werden.

Ergebnisse

Die *Bacillus subtilis*-Mutante *ts* 33-6 (MACH u. ENGELBRECHT 1970) ist durch einen temperaturbedingten Teilungsmodus charakterisiert. Sie teilt sich unter permissiven Bedingungen (30 °C) normal und bildet die für *B. subtilis* typischen

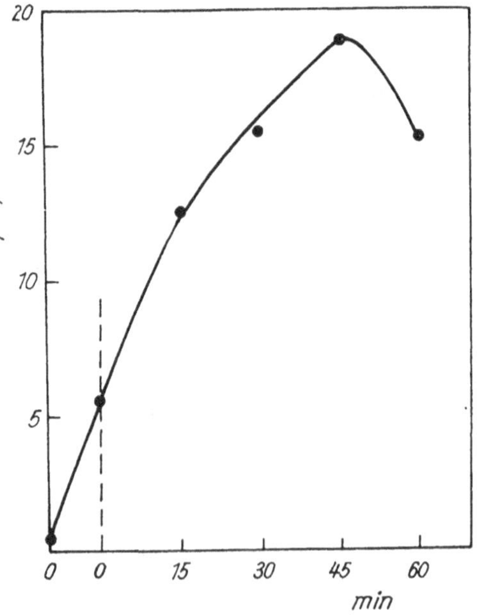

Abb. 1. DNS-Synthese der Mutante *ts* 33-6 im Verlaufe der Keimung bei 45 °C und während der Filamententwicklung bei 46 °C.
Die Sporenkeimung sowie die weitere Kultivierung des Organismus erfolgte im angereicherten Minimalmedium nach COPELAND u. MARMUR (1968), modifiziert. Das Medium enthielt außerdem die keimungsstimulierenden Substanzen L-Alanin und Adenosin.
Radioaktive Konzentration: 0,7 µCi/ml
Die Komplettierung des Mediums durch 0,2% Maisquellwasser erfolgte erst nach Abschluß der Keimung

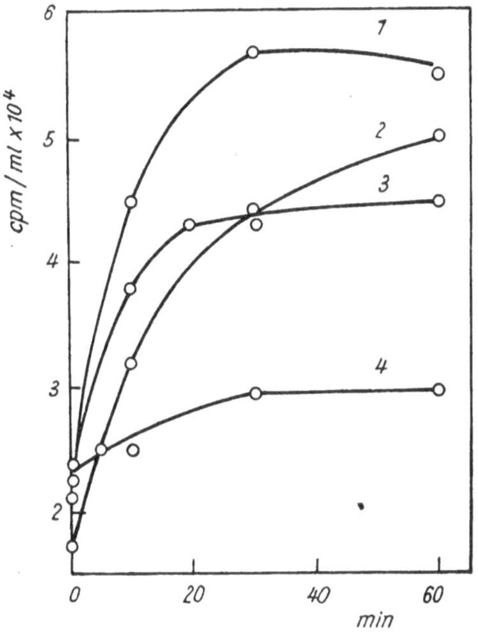

Abb. 2. Inkorporation von ^{3}H-Thymidin (5 µCi/ml) in die Mutante *ts* 33-6 bei 46 °C in Gegenwart von Chloramphenicol.
1 = Kontrollansatz ohne Chloramphenicol, 2 = 10 µg Chloramphenicol/ml, 3 = 20 µg Chloramphenicol/ml, 4 = 20 µg Chloramphenicol/ml 15 min vor Abschluß der Keimung zugesetzt; Keimungsdauer: 30 min

Kurzstäbchen. Nach Temperaturerhöhung auf 46 °C verliert sie die Teilungsfähigkeit und wächst zu filamentösen Formen aus.

Diese filamentösen Formen lassen sich durch Temperaturerniedrigung auf 30 °C wieder zur Teilung anregen und zerfallen nach einer gewissen Zeit mehr oder weniger vollständig.

Vergleicht man die DNS-Synthese bei 30 °C wachsender Zellen und einer filamentösen 46°-Kultur, so lassen sich Unterschiede feststellen (Mach *et al.* 1977). Bei 46 °C ist die Inkorporation vermindert. Besonders intensiv läuft die DNS-Synthese unter restriktiven Bedingungen nur in den ersten 30 min nach dem Transfer der bei 45 °C gekeimten Sporen zu 46 °C ab. Danach ist die Aktivität nur noch gering und kommt zwischen 45 und 60 min vollständig zum Erliegen. Dagegen wird bei permissiver Temperatur (30 °C) auch über diesen Zeitpunkt hinaus noch DNS-synthetisiert. — Die DNS-Synthese, die während der Keimung der Sporen bei 45 °C begonnen wird, kann auch nach Transfer der gekeimten Sporen zu 46 °C, also auch unter restriktiven Temperaturbedingungen, eine gewisse Zeit lang weitergeführt werden (Abb. 1).

Da für den Start (Initiation) der Replikation offensichtlich Proteinbiosynthese nötig ist (Lark 1966, Ward u. Glaser 1969), kann durch den Einsatz von Chloramphenicol abgegrenzt werden, ob die nach Temperaturerhöhung auf 46 °C stattfindende DNS-Synthese unter restriktiven Bedingungen völlig neu initiiert wird oder nicht.

Eine geringe DNS-Synthese der Mutante *ts* 33-6 bei 46 °C unter dem Einfluß von Chloramphenicol deutet darauf hin, daß bestimmte, nicht näher bekannte Prozesse, die die Voraussetzung für die Initiation der Replikation sind, schon während der Keimung der Sporen ablaufen können (Abb. 2.).

Unter dem Einfluß von Chloramphenicol werden jedoch nie die gleichen Inkorporationsmaxima erreicht wie in der unbehandelten Kontrolle. Als Ausdruck der allgemeinen Stoffwechselschädigung durch den Proteinsyntheseinhibitor bleibt die Filamententwicklung aus. Setzt man den keimenden Sporen eine gewisse Zeit vor dem Wechsel zu 46 °C Chloramphenicol zu, so führt diese Schädigung, wie Kurve 4 in Abbildung 2 zeigt, zu einer drastischen Reduktion der Fähigkeit zur DNS-Synthese unter restriktiven Bedingungen. Durch diesen Eingriff wird der Keimungsprozeß derart gestört, daß sich das Potential, das den Syntheseverlauf, wie er in den Kurven 2 und 3 der Abbildung 2 zu sehen ist, nicht mehr aufbauen kann. Voraussetzung für DNS-Replikation unter restriktiven Bedingungen (vgl. Abb. 1) ist ein unter diesen Umständen funktionsfähiger Replikationsmechanismus, d. h. die wichtigsten Enzyme für diesem Vorgang müssen intakt sein.

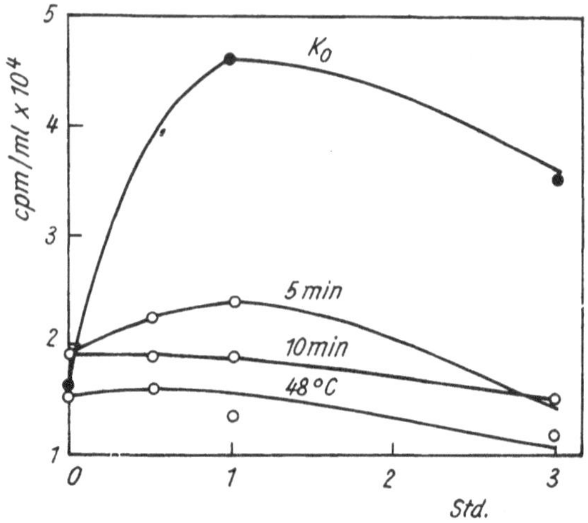

Abb. 3. Bei 45 °C gekeimte, radioaktiv markierte Sporen wurden nach Abschluß der Keimung in den Versuchsansatz, der ebenfalls ^{3}H-Thymidin (5 μCi/ml) enthielt, überimpft. Dann erfolgte die unterschiedlich lange Vorinkubation bei 48 °C (5 bzw. 10 min) und anschließend der Transfer zu 46 °C. *Ko* = Kontrolle 46 °C ohne Vorinkubation; 48 °C = Kontrollansatz, der bei 48 °C verbleibt

Zur Untersuchung der Stabilität des Replikationsmechanismus während der Filamentbildung bot sich folgende Beobachtung an: Überimpft man bei 45 °C gekeimte Sporen der Mutante *ts* 33-6 zu 48 °C und bringt sie nach einer gewissen Zeit (z. B. 5 oder 10 min) zurück zu 46 °C, so ist die Filamentbildung nicht wesentlich beeinträchtigt.

Von Interesse war dabei nicht nur die Fähigkeit gekeimter Sporen der Mutante *ts* 33-6, nach kurzem Aufenthalt bei 48 °C und anschließendem shift down zu 46 °C wieder zu Filamenten auszuwachsen, sondern in erster Linie die Frage, wie sich diese Vorinkubation auf die DNS-Synthese bei 46 °C auswirkt. Das Resultat zeigt Abbildung 3.

Da die optimale Temperatur zur Induktion von Filamenten bei etwa 46,5 °C liegt, genügt offenbar schon ein 5 bzw. 10minütiger Aufenthalt bei einer um *ca.* 2 °C erhöhten Temperatur, um den Replikationsapparat außer Betrieb zu setzen und den gegenüber der Kontrolle völlig veränderten Inkorporationsverlauf zu bewirken.

Die häufig beobachtete Wiederaufnahme der Zellteilung nach Entfernung der teilungshemmenden Bedingungen konnte auch für die Mutante *ts* 33-6 nachgewiesen werden. 150—180 min alte wachstumsaktive Filamente sind in der Lage, nach Temperatur-shift von 46 °C zu 30 °C die Zellteilung wieder aufzunehmen (MACH u. ENGELBRECHT 1970). Die Filamente zerfallen mehr oder weniger vollständig in Kurzstäbchen. Diese Teilungsinduktion ist mit DNS-Synthese verbunden (Abb. 4).

Auf Grund der physiologischen Bedingungen und durch die Unfähigkeit des Organismus zur weiteren Teilung unter restriktiven Temperaturverhältnissen, läßt sich in der bei 46 °C verbleibenden Kontrollkultur keine gesteigerte DNS-Syntheseaktivität nachweisen.

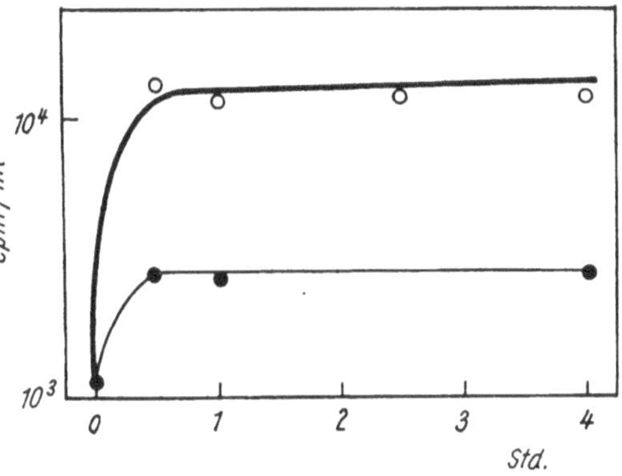

Abb. 4. DNS-Synthese der Mutante *ts* 33-6 bei der Teilungsinduktion durch Temperaturwechsel. 3 Std. alte Filamente wurden von 46 °C zu 30 °C umgesetzt (Induktion ○). Ein bei 46 °C verbleibender Ansatz diente als Kontrolle (●). t_0 kennzeichnet den Zeitpunkt der Zugabe des radioaktiven Thymidins (5 μCi/ml) und des shift down eines Ansatzes von 46 °C zu 30 °C

Die Teilungsinduktion ist nicht nur mit DNS-Syntheseaktivität verbunden. SÜSS u. MACH (1970) fanden in der Phase vor der Teilung der Filamente nach shift down einen erheblichen Anstieg der Syntheserate der Nucleoproteide. Außerdem konnte durch derartige Temperaturwechselexperimente vor dem Zerfall der Filamente ein Anstieg der Aminoacyl-tRNS Synthetasenaktivität nachgewiesen werden (SÜSS *et al.* 1972).

Auf Grund der in Abbildung 4 dargestellten Verhältnisse galt es zu klären, wovon die nach der Temperaturerniedrigung gemessene DNS-Synthese abhängt oder wodurch sie beeinflußt werden kann.

Untersuchungen über den Zusammenhang zwischen Proteinbiosynthese und DNS-Metabolismus nach Temperatur-shift ermöglichen Rückschlüsse über den Start, d. h. die Initiation der nach dem Temperaturwechsel stattfindenden DNS-Synthese. Wenn nach Temperatur-shift mit nachfolgender Teilung der Filamente, wie das bei der Mutante *ts* 33-6 der Fall ist, völlige Neuinitiation der Replikation vorläge, so müßte die Bildung des „Starterproteins" durch Zusatz des spezifischen Proteinsynthesehemmstoffes Chloramphenicol blockiert bzw. stark gehemmt sein. Die Folge davon wäre eine blockierte zumindest jedoch verzögert ablaufende DNS-Synthese. Die DNS-Synthese wird jedoch durch das Antibioticum nicht beeinflußt (Abb. 5).

Der Zusatz von Mitomycin C, einem spezifischen DNS-Syntheseinhibitor, beeinflußt, wie die Resultate verschiedener Experimente zeigten, jedoch konzentrationsabhängig die DNS-Synthese nach Temperaturerniedrigung (Abb. 6).

Das Antibioticum hat zunächst keinen Einfluß auf die RNS- und Proteinsynthese (SHIBA *et al.* 1959), sondern führt zu Doppelstrangbrüchen und blockiert somit die DNS-Replikation. Da neben Proteinsynthese-abhängigen Schritten bei der Initiation der DNS-Replikation auch die RNS-Synthese eine Rolle spielt, kommt dem

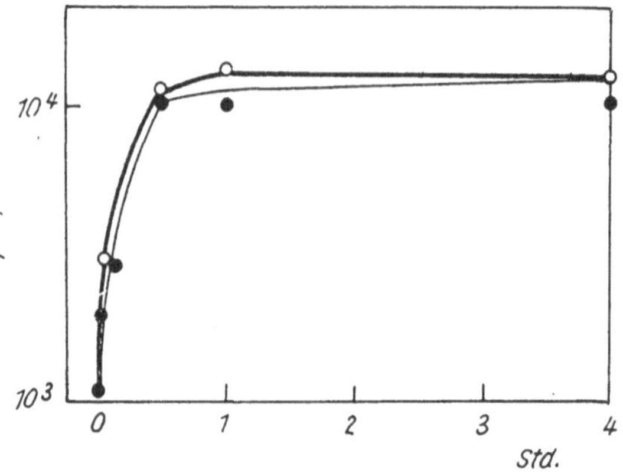

Abb. 5. DNS-Synthese der Mutante *ts* 33-6 nach temperaturbedingter Teilungsinduktion unter Chloramphenicoleinwirkung.
Einem Ansatz von 3 Std. alten wachstumsaktiven Filamenten wurden bei Temperaturwechsel von 46 °C zu 30 °C ^{3}H-Thymidin (5 µCi/ml) und Chloramphenicol (10 µg/ml) zugesetzt (○). Der Kontrollansatz (●) enthielt kein Antibioticum

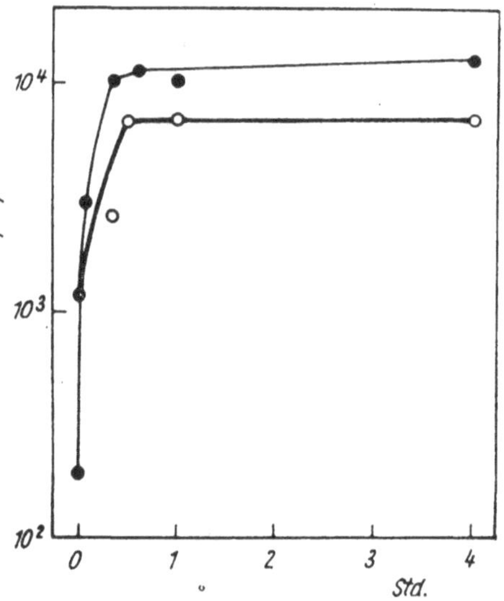

Abb. 6

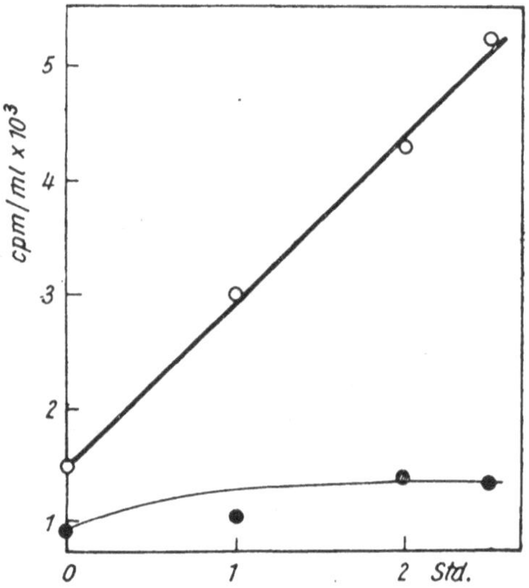

Abb. 7

Abb. 6. DNS-Synthese der Mutante *ts* 33-6 bei temperaturbedingter Teilungsinduktion in Gegenwart von Mitomycin C (○) 3 Std. alte wachstumsaktive Filamente wurden von 46 °C zu 30 °C umgesetzt. Die Zugabe von ^{3}H-Thymidin (5 µCi/ml) und Mitomycin C (0,5 g/ml) erfolgte zum Zeitpunkt des shift down.
Der Kontrollansatz (●) enthielt kein Antibioticum

Abb. 7. Vergleich von DNS- (○) und RNS-Synthese (●) der Mutante *ts* 33-6 nach Teilungsinduktion durch Temperatur in aminosäurefreiem Minimalmedium.
150 min alte, in angereichertem und durch 0,2% Maisquellwasser ergänztem Minimalmedium nach COPELAND u. MARMUR (1968) bei 46 °C in Gegenwart von ^{3}H-Thymidin (0,7 µCi/ml) bzw. ^{3}H-Uracil (0,38 µCi/ml) kultivierte Filamente, wurden abzentrifugiert und in aminosäurefreiem Minimalmedium, das ^{3}H-Thymidin (0,7 µCi/ml) bzw. ^{3}H-Uracil (0,37 µCi/ml) enthielt, resuspendiert

Vergleich von DNS und RNS-Metabolismus nach shift down von Filamenten der Mutante *ts* 33-6 im Sinne der eingangs erwähnten Fragestellung Bedeutung zu. Im Rahmen dieser Untersuchungen läßt sich die Auxotrophie des Organismus für Valin, Leucin, Isoleucin und Glutamin ausnutzen.

Das Nährmedium, in dem nach dem Temperaturwechsel die weitere Kultivierung erfolgte, enthielt weder die Aminosäuren, für die die Mutante auxotroph ist, noch irgendwelche anderen Aminosäuren.

Für die kontinuierlich ablaufende RNS-Synthese ist das Vorhandensein aller Aminosäuren Voraussetzung (Lark *et al.* 1963). Bei Aminosäurehunger, oder wenn Aminosäuren nicht aktiviert werden können, ist die RNS-Neusynthese stark reduziert (Gallant u. Margason 1972).

Die Filamente der Mutante *ts* 33-6 sind nach Transter in aminosäurefreies Medium mit gleichzeitigem Temperaturwechsel in der Lage, DNS zu synthetisieren, obwohl die RNS-Synthese signifikant beeinflußt wird (Abb. 7).

Diskussion

Eine allgemeine Klassifizierung temperatursensitiver Mutanten (Hirota *et al.* 1968) unter Berücksichtigung ihrer DNS-Replikation nach Temperaturerhöhung ergibt zwei voneinander unterschiedliche Gruppen:
a) Mutanten, deren DNS-Synthese nach Temperaturerhöhung sofort blockiert wird
b) Mutanten, deren DNS-Synthese unter nicht-permissiven Bedingungen noch eine gewisse Zeit lang weitergeht.

Die Ursache für das unterschiedliche Verhalten läßt sich folgendermaßen interpretieren (Spratt u. Rowbury 1970):
Während bei den Mutanten der Kategorie *a*) der Schaden durch einen raschen Angriff auf die für die Replikation wichtigen Enzyme verursacht wird, gibt es für die Mutanten der Gruppe *b*) zwei Möglichkeiten der schädigenden Beeinflussung.

Erstens könnte die Replikationsrunde beendet werden, eine neue jedoch nicht mehr gestartet werden. In diesem Falle tritt die Schädigung auf der Stufe der Initiation der Replikation auf. Als zweite Möglichkeit muß eine langsame Inaktivierung wichtiger Enzyme erwähnt werden, ohne daß dabei die Initiation beeinflußt wird.

Während der Keimung der Sporen der Mutante *ts* 33-6 sowohl bei 30 °C als auch bei 45 °C wird, wenn auch mit unterschiedlicher Kinetik, DNS synthetisiert.

Mendelson (1969) konnte ebenfalls während der Keimung von *B. subtilis*-Sporen Thymidininkorporation feststellen und die Verteilung des neusynthetisierten Materials im Verlaufe des weiteren Wachstums der aus den Sporen hervorgegangenen Zellen nachweisen. Die während der relativ schnellen Keimung bei 45 °C begonnene DNS-Replikation kann von der Mutante *ts* 33-6 auch bei nichtpermissiver Temperatur weitergeführt werden. Für die DNS-Synthese, die unter restriktiven Bedingungen während der Filamentbildung ablaufen kann, ist der physiologische Zustand der gekeimten Sporen zum Zeitpunkt des shift up von entscheidender Bedeutung.

Eine besondere Rolle im Regulationsmechanismus der Zellteilung spielt die Initiation der Replikation. Der Prozeß der Initiation erfordert sowohl die Proteinsynthese (Ward u. Glaser 1969) als auch Prozesse, an denen die RNS-Synthese maßgeblichen Anteil hat. Dieser letzte, zum Zeitpunkt der Initiation ablaufende Schritt, läßt sich durch das Antibioticum Rifamycin, das die DNS-abhängige RNS-Polymerase hemmt, blockieren. Der Mechanismus über den die RNS-Synthese an der Initiation beteiligt ist, ist noch unklar (Hanna u. Carl 1975). In einer von Lark (1972) vorgeschlagenen Arbeitshypothese wird eine direkte Beteiligung einer stabilen RNS an der DNS-Replikation angenommen, wobei rund um einen „RNS-core“ angeordnete Proteine einen sogenannten Replikationskomplex bilden.

Untersuchungen zur Umwandlung von Einstrang-DNS des Phagen $\Phi \times 174$ in die replikative Doppelstrangform wiesen auf die Bildung einer kurzen RNS-Kette (primer) hin (SCHEKMAN *et al.* 1972).

Die Initiation der DNS-Replikation durch einen RNS-primer-Mechanismus ist offenbar von weitreichender Bedeutung, da ähnliche Beobachtungen auch bei der Replikation des Phagen M 13 bzw. der Synthese colicinogener Faktoren gemacht wurden.

Die Tatsache, daß während der Keimung der Sporen Prozesse der Initiation der Replikation ablaufen, ermöglicht eine gewisse DNS-Synthese bei 46 °C auch unter dem Einfluß von Chloramphenicol. Über ähnliche Resultate an keimenden Sporen von *B. subtilis* unter Einwirkung dieses Proteinsyntheseinhibitors wurde schon 1965 von YOSHIKAWA berichtet. Die Mutante *ts* 33-6 verfügt unter restriktiven Bedingungen über einen funktionsfähigen Replikationsmechanismus.

Einen Eindruck von der Thermolabilität dieses Apparates vermittelt die Tatsache, daß schon eine kurzzeitige Temperaturerhöhung um etwa 2 °C zu völlig veränderten Verhältnissen für die DNS-Replikation führen kann.

Diese kurzzeitige Erhöhung der Temperatur inaktiviert offenbar die für die Replikation wichtigen Enzyme, die ohne diese Vorinkubation sonst unter restriktiven Bedingungen intakt sind und die nachweisbare DNS-Synthese während der Filamentbildung ermöglichen.

Der Vorinkubationseffekt beeinflußt spezifisch nur die DNS-Synthese, nicht aber den RNS-Metabolismus und die Proteinbiosynthese (SCHRÖDER, unveröffentlicht). Dadurch wird natürlich auch die Filamentbildung nach Vorinkubation gekeimter Sporen nicht behindert.

Abschließend läßt sich feststellen, daß im Verlauf der DNS-Synthese während der Filamentbildung bei 46 °C die vor dem shift up bei der Keimung der Sporen bei 45 °C begonnene Replikation offensichtlich beendet wird. Der dazu benötigte Enzymmechanismus ist, wie die Inaktivierungsversuche mit Hilfe der Vorinkubation bei 48 °C zeigen (Abb. 3), im Normalfall vorhanden.

Ordnet man die Filamente der Mutante *ts* 33-6 nach dem Verhalten ihrer DNS-Replikation in das eingangs dargestellte Schema ein, so gehören sie in die Kategorie *b*), bei denen die DNS-Synthese nach shift up noch eine gewisse Zeit weitergeht.

Aus den Temperaturwechselexperimenten (shift down wachstumsaktiver Filamente von 46 °C zu 30 °C) läßt sich ableiten, daß die nach dem shift down zu bebeobachtende DNS-Synthese keine Proteinsynthese zur Neuinitiation benötigt.

Ebenso laufen unter restriktiven Bedingungen RNS-abhängige Prozesse ab, die die Replikation ohne gleichzeitige RNS-Synthese im aminosäurefreiem Medium gestatten. Der direkte Beweis für die tatsächlich vor sich gehende Replikation ist in der Blockierung der DNS-Synthese durch Mitomycin C zu sehen.

Um die Resultate nach Teilungsinduktion durch Temperatur zu interpretieren, muß man annehmen, daß auch unter restriktiven Bedingungen während der Filamentbildung Voraussetzungen zur Initiation der DNS-Replikation geschaffen werden können.

An anderen Mutanten gewonnene Untersuchungsbefunde (KUEMPEL 1969, BEYERSMANN *et al.* 1971, KOUKALOVA *et al.* 1973) wiesen auf die Möglichkeit der Schaffung eines „Initiationspotentials“ unter restriktiven Bedingungen hin. Der Aufbau eines Initiationspotentials unter Einwirkung erhöhter Temperatur erfordert, wie KOUKALOVA *et al.* (1973) berichten, das Vorhandensein aller Aminosäuren sowie das Funktionieren der Proteinbiosynthese. Eine von diesen Autoren beschriebene Mutante von *Lactobacillus acidophilus* zeigt unter restriktiven Bedingungen zweiphasige

DNS-Synthese. Für die Initiation der zweiten Phase ist unter diesen Bedingungen die Synthese von Proteinkomponenten nötig und auch möglich.

Da bei der Mutante *ts* 33-6 RNS-und Proteinsynthese unter restriktiven Bedingungen normal ablaufen, ist der Aufbau eines solchen „Initiationspotentials" z. B. in inkompletter oder inaktiver Form durchaus möglich. Nach Temperaturerniedrigung und gleichzeitiger Aktivierung dieses „Potentials" könnte dann die DNS-Synthese stattfinden, ohne daß Protein- und RNS-Synthese ablaufen müßten.

Die temperaturbedingte Schädigung der Mutante *ts* 33-6 trifft offensichtlich nicht die eigentliche DNS-Replikation während der Filamentbildung. Dagegen werden Synthesen, die für die Septenbildung nötig sind, durch die Erhöhung der Temperatur stark gehemmt (Mach u. Mach, in Vorbereitung).

Es ist aber ebenso nicht absolut auszuschließen, daß auch Syntheseschritte, die für die erneute Initiation der DNS-Replikation erforderlich sind, nicht mehr vollständig ablaufen können und daher die weitere DNS-Synthese, wie sie unter permissiven Bedingungen zu beobachten ist, unmöglich machen.

Für die gute technische Assistenz sind wir Frau G. Fandry sehr zu Dank verpflichtet.

Literatur

Bazill, G. W. and Retieff, Y., 1969. Temperature-sensitive DNA synthesis in a mutant of *Bacillus subtilis*. J. gen. Microbiol., **56**, 87–97.

Beyersmann, D., Schlicht, M. and Schuster, H., 1971. Temperature sensitive initiation of DNA replication in a mutant of *Escherichia coli* K12. Molec. gen. Genetics, **111**, 145–158.

Copeland, J. C. and Marmur, J., 1968. Identification of conserved genetic functions in *Bacillus* by use of temperature-sensitive mutants. Bacteriol. Rev., **32**, 302–312.

Gallant, J. and Margason, G., 1972. Amino acid control of messenger ribonucleic acid synthesis in *Bacillus subtilis*. J. biol. Chemistry, **247**, 2289–2294.

Hanna, M. H. and Carl, P. L., 1975. Reinitiation of deoxyribonucleic acid synthesis by deoxyribonucleic acid initiation mutants of *Escherichia* coli: Role of ribonucleic acid synthesis and cell division. J. Bacteriol., **121**, 219–226.

Hirota, Y., Ryter, A. and Jacob, F., 1968. Thermosensitive mutants of *Escherichia coli* affected in the process of DNA-synthesis and cellular division. Cold Spring Harbour Sympos. quantitat. Biol., **33**, 677–693.

Koukalova, B., Kuhrova, V. and Soska, J., 1973. A mutant of *Lactobacillus acidophilus* with temperature-sensitive regulation of DNA-synthesis. Folia microbiol., **18**, 7–16.

Kuempel, P., 1969. Temperature-sensitive initiation of chromosome replication in a mutant of *Escherichia coli*. J. Bacteriol., **100**, 1302–1310.

Lark, K. G., Repko, T. and Hoffmann, E. J., 1963. The effect of amino acid depriviation on subsequent deoxyribonucleic acid replication. Biochim. biophysica Acta, **76**, 9–24.

Lark, K. G., 1966. Regulation of chromosome replication and segregation in bacteria. Bacteriol. Rev., **30**, 3–32.

Lark, K. G., 1972. Evidence for the direct involvement of RNA in the initiation of DNA-replication in *Escherichia coli* 15 T⁻. J. molecular Biol., **64**, 47–60.

Mach, F. und Engelbrecht, H., 1970. Isolation und erste Charakterisierung einer temperatursensitiven, filamentösen Mutante von *Bacillus subtilis*. Z. Allg. Mikrobiol. , **10**, 383–395.

Mach, Hiltraut *et al.*, 1977/78. Umfassende Charakterisierung einer temperatursensitiven filamentösen Mutante von *B. subtilis* SB19 (russ.). Mikrobiologia, im Druck.

Mendelson, N. H. and Gross, J. D., 1967. Characterization of a temperature-sensitive mutant of *Bacillus subtilis* defective in DNA-replication. J. Bacteriol., **94**, 1603–1608.

Mendelson, N. H., 1969. Nuclear segregation without DNA-replication in *Bacillus subtilis*. Biochim. biophysica Acta, **190**, 132–138.

Schekman, R. *et al.* 1972. Initiatiation of DNA-synthesis: synthesis of *φ*X 174 replicative form requires RNA synthesis resistant to rifampicin. Proc. nat. Acad. Sci., **69**, 2691–2695.

Shiba, S., Terawaki, A. and Kawabata, J., 1959. Selective inhibition of formation of deoxyribonucleic acid in *Escherichia coli* by mitomycin C. Nature, **183**, 1056–1057.

Spratt, B. G. and Rowbury, R. J., 1970. A mutant of the initiation of DNA synthesis in *Salmonella typhimurium*. J. gen. Microbiol., **64**, 127–138.

SÜSS, J., 1970. Diplomarbeit Sektion Biologie der Ernst-Moritz-Arndt-Universität Greifswald.
SÜSS, J., MACH, HILTRAUT und MACH, F., 1972. Aminoacyl-tRNS Synthetasen und Zellteilung bei einer temperatursensitiven filamentösen Mutante von *Bacillus subtilis* SB 19. Z. Allg. Mikrobiol., **12**, 161–168.
TORRIANI, A. and LEVINTHAL, C., 1967. Ordered synthesis of proteins during outgrowth of spores of *Bacillus cereus*. J. Bacteriol., **94**, 176–183.
URCROFT, P., DYSON, H. 4. and WAKE, R. G., 1975. Charcteristics of a *Bacillus subtilis* W23 mutant temperature sensitive for initiation of chromosome replication. J. Bacteriol., **121**, 121–127.
WARD, C. B. and GLASER, D. A., 1969. Analysis of the chloramphenicol-sensitive and chloramphenicol resistant steps in the initiation of DNA synthesis in *Escherichia coli* B/r. Proc. nat. Acad. Sci., **64**, 905–912.
YOSHIKAWA, H., 1965. DNA synthesis during germination of *Bacillus subtilis* spores. Proc. nat. Acad. Sci., **53**, 1476–1483.

Anschrift: Dr. D. SCHRÖDER
Sektion Biologie, WB Allgemeine Mikrobiologie
der Ernst-Moritz-Arndt-Universität
DDR 22 Greifswald, Jahnstr. 15

Zeitschrift für Allgemeine Mikrobiologie	17	7	1977	559–568

(Akademie der Wissenschaften der DDR, Forschungszentrum für Molekularbiologie und Medizin, Zentralinstitut für Mikrobiologie und experimentelle Therapie, Jena, Direktor: Prof. Dr. U. TAUBENECK)

Einbau von Thymidin in die DNS von Actinomyceten

I. Der Einbau exogenen Thymidins in die DNS von *Thermoactinomyces vulgaris*

G. STROHBACH and SIGRID KRETSCHMER

(Eingegangen am 24. 8. 1976)

The incorporation of exogenous thymidine and thymine into acid-insoluble material of *Thermoactinomyces vulgaris* has been studied during germination and subsequent growth.

Thymine is not incorporated. The incorporation of thymidine stops after a short time due to the rapid breakdown of thymidine to thymine and deoxyribose-1-phosphate by the inducible thymidine phosphorylase.

Deoxyadenosine enhances the incorporation of thymidine as well as of thymine and prolongs the time of uptake. Uridine stimulates only the incorporation of thymidine but not of thymine. These effects can be explained by the function of these substances within the salvage pathway. Deoxyadenosine acts as donor of deoxyribosyl groups being necessary for the conversion of thymine to thymidine by thymidine phosphorylase and uridine inhibits thymidine phosphorylase, and thereby it prevents the degradation of thymidine to thymine.

Thymidine is incorporated into alkali-, RNase- and protease-stable, hot TCA-soluble and DNase-sensitive material. That means that the cellular DNA of *T. vulgaris* can be specifically labelled by radioactive thymidine in the presence of deoxyadenosine and uridine, respectively.

Die Markierung der DNS mit ^{14}C- oder ^{3}H-Thymidin bzw. Thymin ist eine häufig angewandte Methode bei biochemischen, autoradiographischen oder kinetischen Untersuchungen des DNS-Stoffwechsels. Es wird dabei davon ausgegangen, daß Thymidin nur in die DNS eingebaut werden kann, und daß der Nachweis der Radioaktivität relativ einfach, schnell und selbst in Spuren noch exakt ist.

Voraussetzung für eine Inkorporation von exogenem Thymidin in die DNS der lebenden Zelle sind dessen Aufnahme in die Zelle und dessen Einschleusung in den zelleigenen DNS-Syntheseweg, in dem Thymidin nur als Mono-, Di- oder Triphosphat vorkommt. Dieser spezielle Aufnahmeweg wird als „salvage pathway" bezeichnet (Zusammenfassung siehe O'DONOVAN u. NEUHARD 1970).

Es gibt Befunde, nach denen Thymidin aber auch ab- oder umgebaut werden kann. Damit besteht bei Verwendung radioaktiv markierter Vorstufen die Möglichkeit, daß außer DNS noch anderes Zellmaterial radioaktiv markiert wird (CLEAVER 1967, SHAFFER *et al.* 1975). Für eine Reihe von Bakterien ist bekannt, daß ein Thymidinabbau verhindert werden kann, wenn der Desoxyribose-Pool erhöht (BOYCE u. SETLOW 1962) oder die enzymatische Aktivität der Thymidinphosphorylase (EC 2.4.2.4) gehemmt wird (BUDMAN u. PARDEE 1967). Dadurch wird ein kontinuierlicher Thymidineinbau in die DNS prototropher Stämme ermöglicht. Einige Organismen können aber Thymidin überhaupt nicht in ihre DNS einbauen, da keine Thymidinkinase (EC 2.7.1.21) vorhanden ist und deshalb das Thymidin nicht in sein Monophosphat überführt werden kann. Das Fehlen von Thymidinkinase wurde beschrieben bei einigen Cyanophyceen (RESTAINO u. FRAMPTON 1975), Hefen, Pilzen und Algen (GRIVELL u. JACKSON 1968).

Untersuchungen über den Einbauweg von markiertem Thymidin in die DNS von Actinomyceten sind uns bisher nicht bekannt.

In der vorliegenden Arbeit wird über Möglichkeiten einer spezifischen Markierung der DNS von *Thermoactinomyces vulgaris* mit ^3H- bzw ^{14}C-Thymidin berichtet.

Material und Methoden

Der Stamm *Thermoactinomyces vulgaris* G 4 wurde aus selbsterhitztem Heu isoliert (KLINGENBERG 1966).

Der Sporulationsagar setzte sich zusammen aus löslicher Stärke 1%, SERVA-Pepton C/HSF 1%, NaCl 0,05%, K_2HPO_4 0,1%, $MgSO_4 \cdot 7H_2O$ 0,03%, $CaCO_3$ 0,001%, $FeSO_4 \cdot 7H_2O$ 0,001% und Agar 1,5%, pH 7,0. Nach 40 Std. Wachstum bei 50 °C wurde die versporte Kultur mit destilliertem Wasser abgeschwemmt. Die Suspension wurde homogenisiert (Homogenisator nach POTTER u. ELVEHJEM), 6mal mit kaltem aqua dest. gewaschen und zentrifugiert. Die Sporensuspension wurde 8 Tage bei Zimmertemperatur und danach bei −4 °C aufbewahrt. Unter diesen Bedingungen konnte über Wochen eine nahezu 100%ige Keimung der Sporen erzielt werden.

Das Kulturmedium bestand aus: Saccharose 1%, Casamino acids 0,6%, NaCl 0,05%, K_2HPO_4 0,1%, $MgSO_4 \cdot 7\,H_2O$ 0,03%, $CaCO_3$ 0,001%, $FeSO_4 \cdot 7H_2O$ 0,001%, pH 7,0.

Die Kultivierung erfolgte in 50-ml-Kölbchen, die bei 50 °C im Wasserbad geschüttelt wurden. Die Gefäße waren silikonsisiert, um unerwünschten Wandbewuchs zu verhindern. 10 ml Medium wurden mit 0,2 ml einer Sporensuspension beimpft. Die Sporenkonzentration in der Kultur betrug etwa $2 \cdot 10^7$/ml.

Folgende Biochemikalien wurden eingesetzt: Desoxyadenosin *p. a.* (SERVA, Heidelberg), Uridin (REANAL, Budapest), Thymidin *p.a.* und Thymin, reinst (SCHUCHARDT, München).

Folgende radioaktiv markierte Verbindungen von ÚVVVR, Prag, wurden verwendet: Thymidin-methyl-^3H (12,6 Ci/mmol), Thymidin-6-^3H (25,9 Ci/mmol), Thymidin-2-^{14}C (43 mCi/mmol), L-Leucin-^{14}C (125 mCi/mmol) und Thymin-2-^{14}C (53 mCi/mmol).

Die radioaktiven Konzentrationen bzw. spezifischen Aktivitäten im Kulturmedium betrugen: Thymidin-^3H: 4 µCi/1 µg/ml; Thymidin-^{14}C: 0,9 µCi/5 µg/ml; Thymin-^{14}C: 0,9 µCi/2,16 µg/ml; L-Leucin-^{14}C: 1,8 µCi/1,12 µg/ml. Die Einstellung erfolgte mit inaktiver Trägersubstanz.

Die Trichloressigsäure-unlösliche Radioaktivität wurde wie folgt bestimmt: Zu festgesetzten Zeiten wurden 2 Parallelproben von jeweils 0,1 ml der Kultur entnommen und auf Filterscheiben (Chromatographiepapier FN7 des VEB FILTRAK, Niederschlag) mit einem Durchmesser von 25 mm gegeben. Die Filterscheiben wurden getrocknet, 3mal in kalter Trichloressigsäure (5%ig), einmal in 96%igem Äthanol und einmal in Aceton gewaschen. Nach dem Trocknen wurden die Scheiben in Meßgläschen überführt und mit 5 ml Szintillationsflüssigkeit (5 g PPO, 0,3 g POPOP in 1 l Toluol) versetzt.

Die Bestimmung der Zählraten erfolgte im Flüssigkeitsszintillations-Spektrometer Modell 3375 der Firma PACKARD oder Modell WALLAC 81000 der Firma LKB.

Zum Nachweis der Spezifität des Einbaus von Thymidin wurden folgende Methoden angewandt:
1. Behandlung mit 5%iger Trichloressigsäure bei 90 °C: Jeweils 3 Parallelproben (0,1 ml) einer mit ^3H-Thymidin gewachsenen Kultur wurden auf FN7-Filterscheiben pipettiert. Die Filterscheiben wurden im Wasserbad 15 min bei 90 °C gehalten, 2mal mit 96%igem Äthanol und einmal mit Aceton gewaschen.

2. Behandlung mit 0,1 N NaOH bei 90 °C: Die Filterscheiben wurden 15 min im Wasserbad von 90 °C mit 0,1 N NaOH behandelt. Danach erfolgte Waschen mit kalter Trichloressigsäure, Äthanol und Aceton.

3. Enzymbehandlung: Aus der Kultur wurden 5 ml entnommen und durch Membranfilter filtriert. Der Rückstand wurde bei 37 °C mit Trispuffer pH 7,2 gewaschen, in Lysozymlösung (5 mg/ml) aufgenommen und 2 Std. bei 37 °C bebrütet. Danach wurde filtriert, gewaschen mit aqua dest. und Trispuffer und in Trispuffer von 37 °C aufgenommen. Es folgte Bebrütung für 2 Std. bei 37 °C mit DNase und 0,02 M $MgCl_2 \cdot 6\,H_2O$, RNase, Chymotrypsin bzw. einer Kontrolle ohne Enzyme. Die Konzentration der Enzyme betrug jeweils 1 mg/ml. Die Enzyme sind in der Abteilung Biochemie unseres Institutes aus Rinderpankreas isoliert worden.

Jeweils 3 Parallelproben von 0,1 ml wurden auf Membranfilter pipettiert, mit 5%iger Trichloressigsäure und Wasser gewaschen, getrocknet, in Szintillator überführt und im Flüssigkeitsszintillations-Spektrometer gemessen.

Die Extinktion wurde mit einem Photometer (ELPHO) des VEB CARL ZEISS Jena bei 470 nm bestimmt.

Ergebnisse

Einbau von Thymidin in Trichloressigsäure-unlösliches Material des prototrophen Stammes

Thymidinauxotrophe Mutanten des Stammes standen nicht zur Verfügung. Die Selektion solcher Mutanten mit der Aminopterinmethode (OKADA *et al.* 1960), die bei verschiedenen Bakterienstämmen erfolgreich angewandt wurde, war bisher ohne Erfolg (STROHBACH, unveröffentlicht).

Deshalb wurde der Einbau von Thymidin in Trichloressigsäure-unlösliches Material des prototrophen Stammes untersucht. Die Kulturbedingungen sind unter Material und Methoden beschrieben. Wie Abbildung 1 zeigt, baut der prototrophe Stamm Thymidin ein. Der Einbau endet jedoch zu einem Zeitpunkt, zu dem der Leucineinbau noch in vollem Gange ist. Der Einbau erfolgt unabhängig davon, ob ringmarkiertes (C-6) oder in der CH_3-Gruppe markiertes Thymidin verwendet wird.

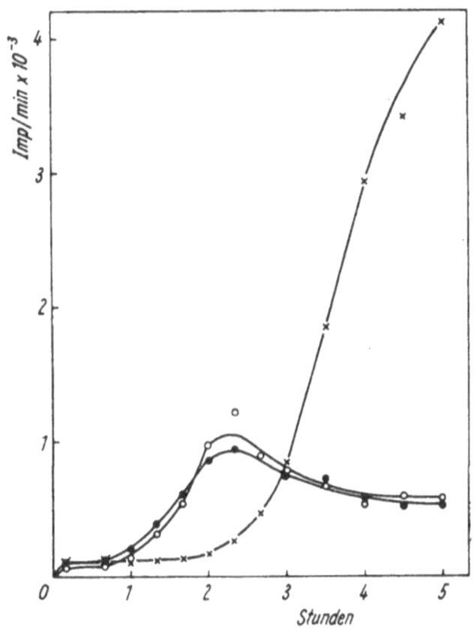

Abb. 1. Thymidin- und Leucineinbau in *Thermoactinomyces vulgaris*.
● Thymidin-methyl-^{3}H, ○ Thymidin-6-^{3}H, × *L*-Leucin-^{14}C

Die Beendigung des Thymidineinbaus zu einem Zeitpunkt, zu dem noch Wachstum erfolgt, wirft die Frage nach den Ursachen des Einbaustopps auf.

Aus Tabelle 1 wird ersichtlich, daß weder eine Erhöhung der Thymidinkonzentration auf das Fünffache bei gleicher spezifischer Aktivität, noch eine Erhöhung der spezifischen Aktivität auf das Fünffache die Einbauzeit verlängern. Lediglich die eingebaute Menge ist abhängig von der Thymidinkonzentration bzw. von der spezifischen Aktivität.

Wie entsprechende Untersuchungen zeigten, ist die Inkorporation auch nicht auf den in Tabelle 1 angegebenen Zeitraum beschränkt. Spätere oder mehrmalige Applikation von Thymidin während der Wachstumsphase führen ebenfalls zu kurzzeitigem Einbau.

Aus diesen Befunden muß der Schluß gezogen werden, daß der Einbau von Thymidin in den prototrophen Stamm unabhängig von der noch ablaufenden DNS-Synthese eingestellt wird.

Tabelle 1

Einbau von Thymidin in Abhängigkeit von der Thymidinkonzentration im Medium bzw. von der spezifischen Aktivität. Die spezifischen Aktivitäten sind in Kultur 1 und 2 gleich. In Kultur 2 ist die Thymidinkonzentration 5mal höher als in Kultur 1. In Kultur 3 ist bei gleicher Thymidinkonzentration im Medium die spezifische Aktivität 5mal höher als in Kultur 1.

R_i = Zählrate [Imp./min]
F_i = Faktor, bezogen auf die zu jeder Entnahmezeit auf 1 normierte Zählrate in der Kultur 1
K = Thymidinkonzentration im Medium [μg Thymidin/ml Kultur]
A = spezifische Aktivität [μCi/μg Thymidin]

Zeit min	R_1 Imp./min	F_1	R_2 Imp./min	F_2	R_3 Imp./min	F_3
10	236	1	974	4,1	1286	5,4
60	744	1	2799	3,8	3394	4,5
90	926	1	3412	3,7	3677	3,9
120	915	1	3402	3,7	3616	3,9
150	794	1	3206	4,0	2897	3,7
180	778	1	2977	3,8	2901	3,7
210	826	1	3062	3,7	2788	3,4
240	853	1	3449	4,1	2809	3,3
K	1 μg/ml		5 μg/ml		1 μg/ml	
A	4 μCi/μg		4 μCi/μg		20 μCi/μg	

Wirkung von Desoxyadenosin bzw. Uridin auf den Thymidineinbau

Desoxyadenosin und Uridin fördern bei verschiedenen Bakterien den Einbau von exogenem Thymidin in die DNS (Boyce u. Setlow 1962, Budman u. Pardee 1967).

Bei *T. vulgaris* ergibt ein Zusatz von 250 μg Desoxyadenosin bzw. Uridin pro ml Kulturlösung eine deutliche Steigerung der eingebauten Aktivität und eine wesentliche Verlängerung der Einbauzeit (Abb. 2). Die Wirkungen von Desoxyadenosin und Uridin sind konzentrationsabhängig. Eine Erhöhung der Konzentrationen über 250 μg/ml bringt keine wesentliche Einbausteigerung, eine deutliche Verringerung der Konzentrationen führt zu vermindertem Einbau. Da Desoxyadenosin und Uridin den Thymidineinbau in gleichem Maße fördern, wurden die meisten weiteren Untersuchungen in Gegenwart von 250 μg Desoxyadenosin pro ml Medium durchgeführt. Aus Untersuchungen an *Alcaligenes faecalis* (Lark 1960), *Agmenellum quadruplicatum* (Ingram u. Fisher 1972) und *Anacystis nidulans* (Restaino u. Frampton 1975) ist bekannt, daß Desoxyadenosin in der in unserem Experiment angewandten Konzentration Wachstum und DNS-Synthese hemmen kann. Für *T. vulgaris* konnte eine solche Hemmwirkung nicht nachgewiesen werden. Es wird im Gegenteil eine leichte Förderung von Wachstum und Leucineinbau (Abb. 2) beobachtet. Die Wirkung von Desoxyadenosin auf die Thymidininkorporation ist allerdings spezifisch. Sie kann nicht durch die geringe Wachstumszunahme verursacht sein.

Die Aktivität, die in Gegenwart von Desoxyadenosin eingebaut wird, ist abhängig von der spezifischen Aktivität des angebotenen Thymidins und bei gleicher spezifischer Aktivität in geringerem Maße auch von der im Medium vorhandenen Thymidinkonzentration (Tab. 2).

Einbau von Thymidin

In den folgenden Versuchen wurde der Einbau der Pyrimidinbase Thymin vergleichend mit dem Einbau des Desoxyribosids Thymidin untersucht. Abbildung 3

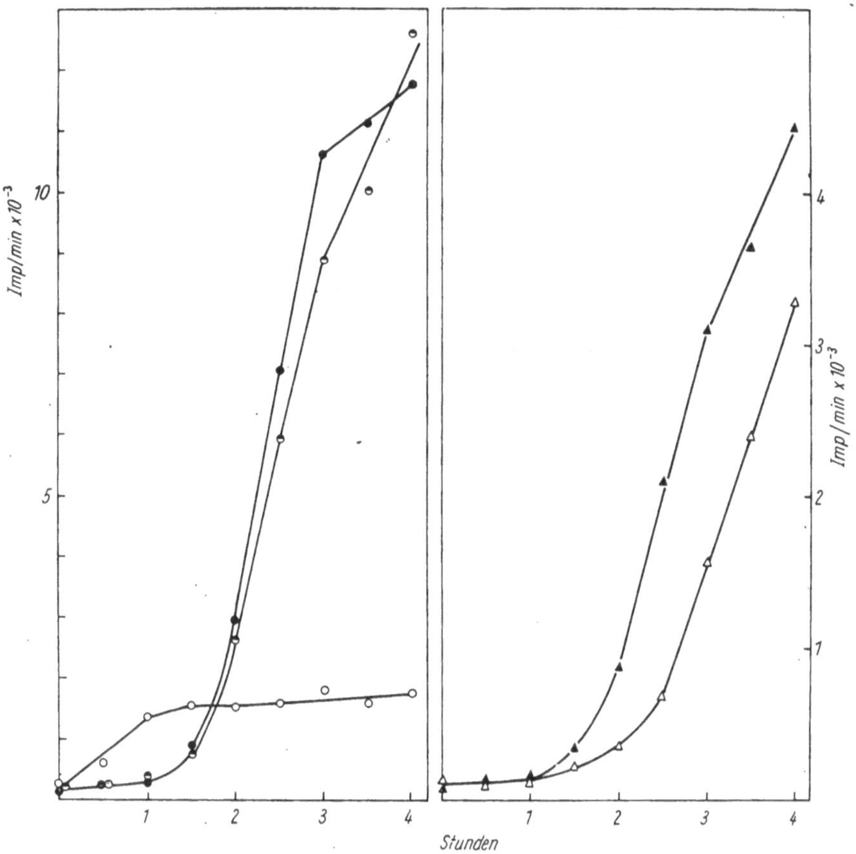

Abb. 2. Einbau von Thymidin-2-^{14}C und *L*-Leucin-^{14}C in Gegenwart von Desoxyadenosin bzw. Uridin.

● Thymidineinbau in Gegenwart von Desoxyadenosin, ◓ Thymidineinbau in Gegenwart von Uridin, ○ Thymidineinbau ohne Zugabe, ▲ Leucineinbau in Gegenwart von Desoxyadenosin. △ Leucineinbau ohne Zugabe

Tabelle 2

Einbau von Thymidin in Abhängigkeit von der Thymidinkonzentration im Medium bzw. von der spezifischen Aktivität in Gegenwart von Desoxyadenosin. Die Bedingungen und Bezeichnungen entsprechen den Angaben zu Tabelle 1

Zeit min	R_1 Imp./min	F_1	R_2 Imp./min	F_2	R_3 Imp./min	F_3
10	204	1	993	4,9	1101	5,0
60	606	1	1902	3,1	3146	5,2
90	2019	1	4056	2,1	10050	4,6
120	7465	1	12345	1,7	32170	4,3
150	20212	1	31712	1,6	100985	5,0
180	31813	1	52964	1,7	151294	4,8
210	38757	1	79805	2,1	228468	5,9
240	37742	1	84327	2,3	236255	6,3
K	1 µg/ml		5 µg/ml		1 µg/ml	
A	4 µCi/µg		4 µCi/µg		20 µCi/µg	

zeigt, daß Thymin nicht eingebaut wird. Durch Zugabe von Desoxyadenosin (250 µg/ml) wird erreicht, daß Thymin in gleicher Weise wie Thymidin inkorporiert wird. Dagegen führt die Applikation von Uridin nicht zu einer Stimulation des Thymineinbaus.

Die relativ hohen Werte bei der Probeentnahme sofort nach Aktivitätszugabe ($t = 0$) werden einer Adsorption zugeschrieben, deren Ursachen noch nicht geklärt sind. Diese Anfangswerte sind bei Thyminapplikation höher als bei Thymidinzugabe.

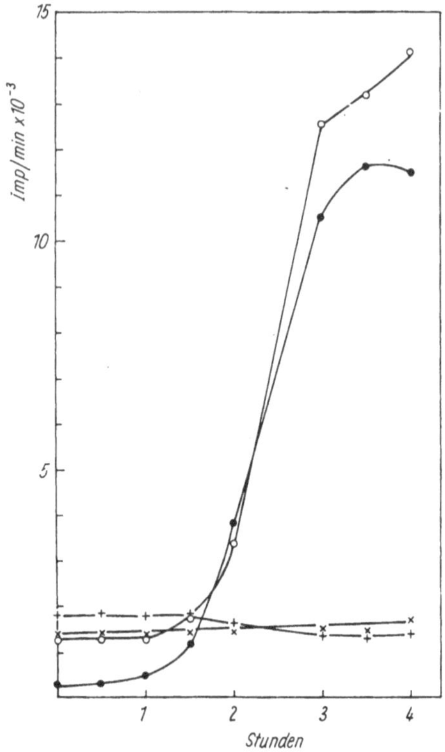

Abb. 3. Einbau von Thymin-2-^{14}C in *Thermoactinomyces vulgaris* und dessen Beeinflussung durch Zugabe von Desoxyadenosin bzw. Uridin.
× Thymineinbau ohne Zugabe, + Thymineinbau in Gegenwart von Uridin, ○ Thymineinbau in Gegenwart von Desoxyadenosin, ● Einbau von Thymidin-2-^{14}C in Gegenwart von Desoxyadenosin

Nachweis der Spezifität des Thymidineinbaus in die DNS

Um den Nachweis zu führen, daß die Aktivität, die in den Trichloressigsäure-unlöslichen Zellbestandteilen vorliegt, spezifisch in die DNS eingebaut wird, wurden Versuche zum hydrolytischen bzw. enzymatischen Abbau der Aktivität durchgeführt. Die Ergebnisse sind in Tabelle 3 dargestellt.

Trichloressigsäure-Behandlung bei 90 °C und DNase-Einwirkung führen zu einem Aktivitätsabbau von mehr als 90%. Dagegen erweist sich die Aktivität als stabil gegenüber NaOH, RNase und Protease.

Der Anstieg der Radioaktivität nach Behandlung mit RNase bzw. Chymotrypsin auf über 100% ist darauf zurückzuführen, daß Substanz abgebaut und damit Schichtdicke und Selbstabsorption verringert werden.

Tabelle 3
Säureunlösliche Radioaktivität nach hydrolytischer bzw. enzymatischer Behandlung

Behandlung	säureunlösliche Radioaktivität (%)
unbehandelte Kontrolle	100
5% TCE, 90 °C, 15 min	7
0,1 NaOH, 90 °C, 15 min	96
DNase, 1 mg/ml, 37 °C, 2 Std.	7
RNase, 1 mg/ml, 37 °C, 2 Std.	111
Chymotrypsin, 1 mg/ml, 37 °C, 2 h	120

Untersuchungen zur Einbaukinetik

In den bisherigen Versuchen wurden unmarkierte Sporen und radioaktiv markiertes Thymidin den Kulturen mit Versuchsbeginn gleichzeitig zugegeben. In den folgenden Versuchen kamen vormarkierte Sporen zum Einsatz.

Sporulationsagar mit 1 µg inaktivem Thymidin, 25 µCi ^{3}H-Thymidin und 250 µg Desoxyadenosin pro ml Agar wurde mit Sporen von *Thermoactinomyces* beimpft. Nach dem Wachsen und Versporen der Kulturen wurden die Sporen geerntet und gewaschen. Die in den Sporen nachgewiesene Aktivität bleibt bei Kultivierung im inaktiven Medium während des Auskeimungs- und Wachstumsprozesses im säureunlöslichen Material gebunden.

Es wurde die Einbaukinetik von ^{3}H-Thymidin in Kulturen verglichen, die mit vormarkierten bzw. unmarkierten Sporen beimpft waren. Die graphische Dar-

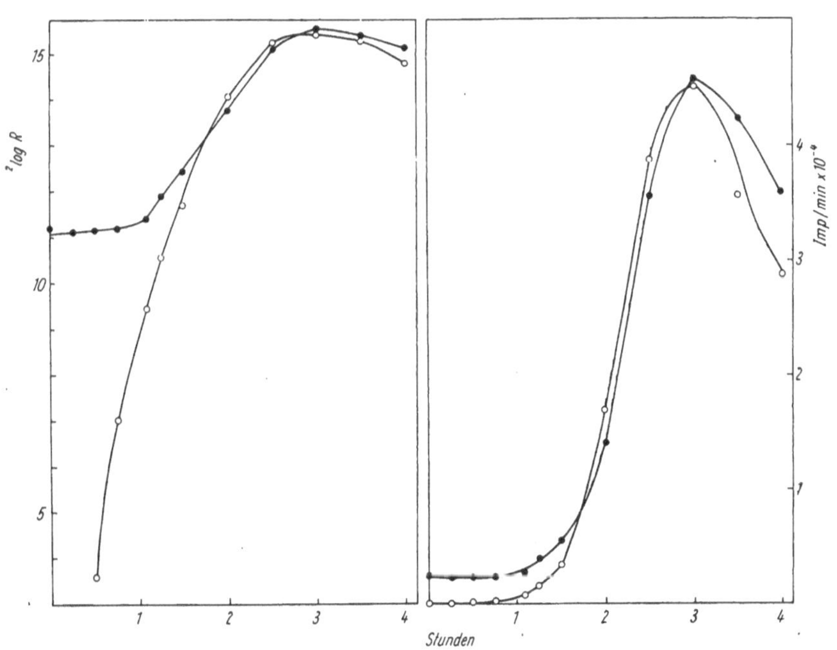

Abb. 4. Einbau von Thymidin-^{3}H in Gegenwart von Desoxyadenosin bei Verwendung vormarkierter bzw. unmarkierter Sporen.

● vormarkierte Sporen, ○ unmarkierte Sporen; R = Zählrate Imp./min; Rechts: lineare Darstellung des Einbauprozesses stellung der Ergebnisse erfolgte halblogarithmisch unter Verwendung des Logarithmus zur Basis 2. Dadurch kann die Verdopplung der eingebauten Radioaktivität unmittelbar abgelesen werden. In den Kulturen mit vormarkierten Sporen wird während der ersten 60 min nur wenig Aktivität eingebaut. Danach vollzieht sich sofort der Übergang zu einem exponentiellen Einbau (Abb. 4 und 5).

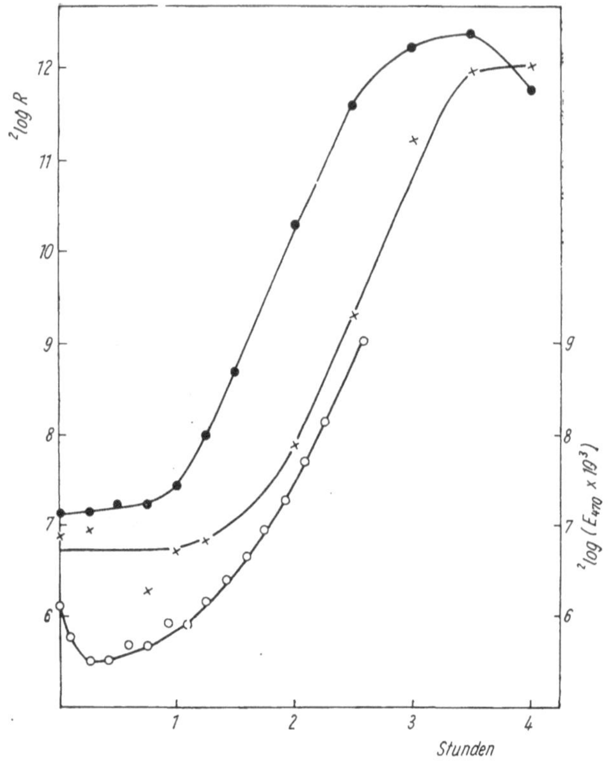

Abb. 5. Einbau von Thymidin-^{3}H, Leucin-^{14}C und optische Dichte (Extinktion) in Kulturen von *Thermoactinomyces vulgaris*. Die eingesetzten Sporen waren mit Thymidin-^{3}H vormarkiert. Die Kulturen enthielten Desoxyadenosin (250 µg/ml).

● Thymidineinbau, × Leucineinbau, ○ Extinktion bei 470 nm; R = Zählrate Imp./min, E = Extinktion

Werden unmarkierte Sporen verwendet, ergibt sich wegen des anfänglich hohen Anteils an unmarkierter DNS eine Einbaukinetik, die sich asymptotisch den Werten der Kultur mit vormarkierten Sporen nähert. Die Einbaukinetik repräsentiert die DNS-Synthesekinetik erst dann, wenn die vorliegende DNS vollständig markiert ist. Wird unmarkiertes Ausgangsmaterial verwendet, ist das nach mehr als 3 DNS-Verdopplungen zu erwarten. Bei Verwendung vormarkierten Impfmaterials wird diese Schwierigkeit umgangen (Smith u. O'Leary 1968). Zum Vergleich erfolgt in Abbildung 4 auch die lineare Darstellung der Meßwerte.

In Abbildung 5 wird die Kinetik der Thymidininkorporation mit der Einbaukinetik von Leucin und einer Wachstumsbestimmung durch Extinktionsmessung verglichen. Die Verdopplungszeiten für alle 3 gemessenen Parameter liegen bei 20 min.

Die relativ großen Streuungen in der lag-Phase des ^{14}C-Leucin-Einbaus sind auf die „spill-over"-Korrekturen der anfänglich sehr niedrigen Zählraten im ^{14}C-Kanal zurückzuführen.

Die Extinktionswerte nehmen während der Sporenquellung ab und steigen erst mit Beginn des Auswachsens des Keimschlauches wieder an. Bei etwa 75 min beginnt die Leucininkorporation und gleichzeitig treten die ersten sichtbaren Keimschläuche auf. Der Einbau von Thymidin beginnt schon früher.

Diskussion

Von *T. vulgaris* konnten bisher keine thyminauxotrophen Mutanten selektiert werden. Die vorliegenden Ergebnisse zeigen aber, daß der prototrophe Stamm in der Lage ist, Thymidin in seine DNS zu inkorporieren. Offenbar wird dafür der für verschiedene Eubakterien beschriebene „salvage pathway" der Thymidinaufnahme benutzt. Diese Feststellung ist wichtig, weil einige Algen, Hefen und Pilze keine Thymidinkinase besitzen und deshalb exogen angebotenes Thymidin nicht in ihre DNS einbauen können.

Die zeitlich begrenzte Aufnahme des Thymidins durch den prototrophen Stamm ist auf einen schnellen Abbau des Thymidins zu Thymin und Desoxyribose-1-phosphat durch die Thymidinphosphorylase zurückzuführen (STROHBACH, in Vorbereitung), wie es auch bei *Escherichia coli* festgestellt werden konnte (RACHMELER *et al.* 1961).

Die Zugabe von Desoxyadenosin führt zur Erhöhung des Desoxyribosidpools, da das Desoxyribosid durch die Wirkung der Purinnucleosidphosphorylase (EC 2.4.2.1) in die Base und Desoxyribose-1-phosphat gespalten wird. Dadurch verschiebt sich das Schwergewicht der Thymidinphosphorylase-Wirkung vom Thymidinabbau in Richtung Thymidinsynthese (BOYCE u. SETLOW 1962).

Die Wirkung von Uridin sehen BUDMAN u. PARDEE (1967) in einer kompetitiven Hemmung der Thymidinphosphorylase. Das angebotene Thymidin wird deshalb nicht abgebaut.

Durch die Zugabe von Desoxyadenosin oder Uridin wird die zeitliche Begrenzung der Inkorporation exogenen Thymidins aufgehoben und ein kontinuierlicher Einbau während der gesamten Wachstumsphase gewährleistet.

Im Gegensatz zum Desoxyribosid wird Thymin wegen des zu geringen Desoxyribosepools der Hyphen nicht eingebaut. Bei Erhöhung des Desoxyribosepools durch Zugabe von Desoxyadenosin wird Thymin in gleichem Maße wie Thymin eingebaut. Damit wird die Funktion des Desoxyadenosins als Desoxyribosedonor bestätigt. Uridin dagegen wirkt als Inhibitor der Thymidinphosphorylase und verhindert die Glykosidierung des Thymins und damit dessen Einschleusung in den „salvage pathway". In Gegenwart von Uridin findet deshalb auch kein Thymineinbau statt. Die Versuche zum Nachweis der Spezifität des Thymidineinbaus in die DNS von *T. vulgaris* zeigen, daß zumindest 90% der Aktivität in der DNS vorliegen. Der Verbleib der restlichen Aktivität konnte experimentell nicht mit Sicherheit nachgewiesen werden. Wir gehen aber davon aus, daß das angebotene Thymidin ausschließlich in die DNS eingebaut wird.

In Gegenwart von Desoxyadenosin repräsentiert die Einbaukinetik des radioaktiv markierten Thymidins offensichtlich auch die DNS-Synthesekinetik, wenn von vormarkierten Sporen ausgegangen wird. Abbildung 4 demonstriert, daß bei der Verwendung unmarkierter Sporen die in halblogarithmischer Darstellung aufgezeichnete Einbaukinetik nicht der DNS-Synthesekinetik entsprechen kann. Dies ist darauf zurückzuführen, daß in der Anfangsphase des Einbaus eine Aktivitätsverdopplung nicht einer DNS-Verdopplung entspricht, weil ein Teil der DNS noch unmarkiert vorliegt. Nach etwa 3 Verdopplungen der Gesamt-DNS ist nahezu alle DNS radioaktiv markiert und erst dann entspricht die Einbaukinetik der DNS-Synthesekinetik.

Ein Vergleich von DNS-Replikation (^{3}H-Thymidin-Inkorporation), Proteinsynthese (^{14}C-Leucin-Einbau) und Wachstum (Extinktionsmessung) zeigt gute Übereinstimmung. Die Verdopplungszeiten liegen bei 20 min.

Der Extinktionsabfall während der Quellung weist auf Dichteveränderungen hin. Das Auftreten sichtbarer Keimschläuche und der Beginn der Leucininkorporation sind nach etwa 75 min gleichzeitig zu registrieren. Die DNS-Replikation beginnt nach den vorliegenden Untersuchungen etwas früher.

Zur Klärung der Zusammenhänge zwischen DNS, RNS- und Proteinsynthese während der Keimungsphase sind weitere Versuche geplant.

Frau G. KASCH danken wir für sorgfältige Versuchsdurchführung.

Literatur

BOYCE, R. P. and SETLOW, R. B., 1962. A simple method of increasing the incorporation of thymidine into the deoxyribonucleic acid of *Escherichia coli*. Biochim. biophysica Acta, **61**, 618–620.

BUDMAN, D. R. and PARDEE, A. B., 1967. Thymidine and thymine incorporation into deoxyribonucleic acid: inhibition and repression by uridine of thymidine phosphorylase of *Escherichia coli*. J. Bacteriol., **94**, 1546–1550.

CLEAVER, J. E., 1967. Thymidine Metabolism and Cell Kinetics. North-Holland Publishing Company Amsterdam.

GRIVELL, A. R. and JACKSON, J. F., 1968. Thymidine kinase: Evidence for its absence from *Neurospora crassa* and some other micro-organisms, and the relevance of this to the specific labelling of deoxyribonucleic acid. J. gen. Microbiol., **54**, 307–317.

INGRAM, L. O. and FISHER, W. D., 1972. Selective inhibition of deoxyribonucleic acid synthesis by 2-deoxyadenosine in the blue-green bacterium *Agmenellum quadruplicatum*. J. Bacteriol., **112**, 170–175.

KLINGENBERG, P., 1966. Taxonomische und physiologische Untersuchungen an thermophilen Streptomyceten. Diplomarbeit. Ernst-Moritz-Arndt-Universität Greifswald.

LARK, K. G., 1960. Studies on the mechanism regulating periodic DNA synthesis in synchronized cultures of *Alcaligenes fecalis*. Biochim. biophysica Acta, **45**, 121–132.

O'DONOVAN, G. A. and NEUHARD, J., 1970. Pyrimidine metabolism in microorganisms. Bacteriol. Rev., **34**, 278–343.

OKADA, T., YANAGASIWA, K. and RYAN, F. J., 1960. Elective production of thymineless mutants. Nature, **188**, 340–341.

RACHMELER, M., GERHART, J. and ROSNER, J., 1961. Limited thymidine uptake in *Escherichia coli* due to an inducible thymidine phosphorylase. Biochim. biophysica Acta, **49**, 222–225.

RESTAINO, L. and FRAMPTON, L. W., 1975. Labeling the deoxyribonucleic acid of *Anacystis nidulans*. J. Bacteriol., **124**, 155–160.

SHAFFER, P. M., HSU, C.-A. and ABBOTT, M. T., 1975. Metabolism of pyrimidine deoxyribonucleosides in *Neurospora crassa*. J. Bacteriol., **121**, 648–655.

SMITH, K. C. and O'LEARY, M. E., 1968. The pitfalls of measuring DNA synthesis kinetics as exemplified in ultraviolet radiation studies. Biochim. biophysica Acta, **169**, 430–438.

Anschrift: Dr. G. STROHBACH
Zentralinstitut für Mikrobiologie und experimentelle Therapie der AdW
DDR 69 Jena, Beuthenbergstr. 11

Zeitschrift für Allgemeine Mikrobiologie	17	7	1977	569–573

Kurze Originalmitteilung

(Akademie der Wissenschaften der DDR, Forschungszentrum für Molekularbiologie und Medizin, Zentralinstitut für Mikrobiologie und experimentelle Therapie, Jena, Direktor: Prof. Dr. U. TAUBENECK)

Photodynamische Wirkung und Reparatur-Inhibition photodynamisch induzierter DNA-Schäden in *Proteus mirabilis* durch das Anthracyclin-Antibioticum Violamycin BI

H.-E. JACOB, W. FLECK und D. STRAUSS

(Eingegangen am 13. 12. 1976)

Violamycin BI ist ein Komplex von Anthracyclin-Antibiotica, der hauptsächlich aus den Trisacchariden der Aglycone Epsilon-Isorhodomycinon, Beta-Rhodomycinon-$(Alpha)_2$-Rhodomycinon und Zeta-Isorhodomycinon zusammengesetzt ist (FLECK *et al.* 1975). Zuckerkomponenten der Trisaccharide sind Rhodosamin, Desoxy-L-fucose und Rhodinose (FLECK *et al.* 1974). Anthracycline sind Inhibitoren der Nucleinsäure-Biosynthese, und der molekulare Wirkungsmechanismus beruht auf einer Interkalation der Anthracyclin-Moleküle zwischen den Basenpaaren der DNA, wodurch die „template"-Aktivität der DNA reduziert wird und Störungen der Replikations- und Transkriptions-Prozesse eintreten (KERSTEN u. KERSTEN 1974, DIMARCO *et al.* 1975). Da es sich bei den Anthracyclinen um rote Pigmente mit Indikatoreigenschaften handelt, sind bereits frühzeitig Untersuchungen zur photodynamischen Wirkung des Daunorubicins durchgeführt worden (SANFILLIPPO *et al.* 1968, VERINI *et al.* 1968, ZUNINO *et al.* 1975). In der vorliegenden Arbeit wird daher geprüft, ob Violamycin BI als neuer Vertreter der Anthracyclin-Antibiotica mit einer vom Daunorubicin abweichenden Aglycon-Struktur und Zuckerzusammensetzung auch eine photodynamische Wirkung bei *Proteus mirabilis* entfalten kann. Darüber hinaus wurde auch untersucht, ob Violamycin BI die Reparatur photodynamisch induzierter DNA-Schäden zu hemmen vermag. Von STÖRL *et al.* (1977) war gezeigt worden, daß Violamycin BI bei Konzentrationen von $5 \cdot 10^{-5}$ M die Excisionsreparatur in UV-bestrahlten Zellen von *Escherichia coli* K12 AB 1157 stören und in unbestrahlten Zellen Einzelstrangbrüche induzieren kann.

Für die Untersuchungen sowohl zur photodynamischen Wirkung als auch zur Reparatur-Inhibition photodynamisch induzierter DNA-Schäden sind Zellen des Bakterienstammes *Proteus mirabilis* VI PG 273 (BÖHME 1967) eingesetzt worden, die zur Reparatur photodynamisch induzierter DNA-Schäden befähigt sind (BÖHME 1968, JACOB 1975). Nach 3stündiger Kulturdauer in Nährmedium (Maltose 10,0 g, Na-Laktat (50%ig) 20,0 ml, K_2HPO_4 1,0 g, Casamino acids (DIFCO) 5,0 g, Lösung A 4,0 ml, Lösung B 2,0 ml, dest. Wasser zu 1 l, pH 7,5; Lösung A: Vitamin B12 1,0 g, Nikotinsäureamid 0,1 g in 100 ml dest. Wasser. Lösung B: $MgSO_4 \cdot 7\,H_2O$ 2,0 g, NaCl 0,1 g, $FeSO_4 \cdot 7\,H_2O$ 0,1 g, $MnSO_4 \cdot 4\,H_2O$ 0,3 g in 50 ml dest. Wasser) wurden die in der logarithmischen Wachstumsphase befindlichen Zellen mittels Zentrifugation vom Nährmedium abgetrennt, 1mal mit isotonischer NaCl-Lösung gewaschen und mit dem der Menge des Nährmediums entsprechenden Volumen Boratpuffer (0,06 M Na-Tetraborat, 0,4 M NaCl, pH 8,9) aufgenommen. Die Belichtung der Bakterienkultur erfolgte im Wasserbad bei 37 °C mit Hilfe einer 150 W Wolfram-Lampe (VEB Berliner Glühlampenwerk) bei einem Abstand zwischen Lampe und Gefäß von 30 cm.

Violamycin BI war in Konzentrationen von $1 \cdot 10^{-4}$ M bis $1 \cdot 10^{-5}$ M den Reaktionsansätzen beigegeben. Als photodynamisch wirksame Kontrollsubstanz diente Methylenblau bei einer Konzentration von $1{,}7 \cdot 10^{-6}$ M. Verschiedene Zeiten nach der Bestrahlung mit sichtbarem Licht wurden aliquote Anteile der Zellsuspension entnommen, nach entsprechender Verdünnung auf Nähragar (Inst. f. Immunpräparate DDR-112 Berlin) aufgespatelt und bei 37 °C inkubiert. 16 Std. danach erfolgte die Titerbestimmung der Zellsuspensionen auf Grund der Zählung koloniebildender Einheiten (KbE).

Zur Untersuchung des Violamycin BI-Einflusses auf die Reparatur photodynamisch induzierter DNA-Schäden wurde die DNA in den Zellen radioaktiv markiert. Die Kultivierung der Bakterien erfolgte in 5 ml Nährmedium in Anwesenheit von ^{3}H-Thymidin (spezifische Aktivität 30 µCi/ml) und Desoxyadenosin (0,25 mg/ml) für 3 Stunden. Die Zellen wurden vom Medium durch Zentrifugation 15000 g für 30 min) oder mit Hilfe von Synpor-Filtern (Größe 6, Poren-⌀ 0,40 µm) abgetrennt. Die in Boratpuffer resuspendierten Zellen wurden mit 1,7 µM Methylenblau versetzt und 8,5 sec belichtet. Während der Reparatur der durch Methylenblau induzierten DNA-Schäden war Violamycin BI dem belüfteten Nährmedium bei 37 °C für die Dauer von 20 min in Konzentrationen von $2{,}5 \cdot 10^{-5}$ M bis $1 \cdot 10^{-4}$ M zugesetzt. Zum Nachweis von Schädigung bzw. Reparatur mußte die DNA isoliert werden. Das geschah durch Lyse der Zellen in einem unmittelbar vor Gebrauch frisch bereiteten 3-Komponenten-Gemisch (8,0 g Saccharose, 0,5 g KCl, 3,7 g EDTA und 0,4 g TRIS in 100 ml aqua dest gelöst, pH 8,0). 0,4 ml dieser Lösung wurden gemischt mit 0,12 ml einer 1%igen Na-N-Laurylsarcosinate-Lösung (Serva-Heidelberg) und 0,28 ml Lysozym (Serva Heidelberg). Die Lysozym-Stammlösung bestand aus 20 mg/2 ml aqua dest. Zu 0,3 ml dieses Gemisches wurden 0,2 ml Bakteriensuspension gegeben und davon 0,2 ml auf den alkalischen Saccharosegradienten (5—20%) langsam in die auflagernde 0,5 M Natronlauge (0,1 ml) einpipettiert. Nach Zentrifugation mit dem SW-39-Rotor einer Spinco E (Beckman — Palo Alto) bei 36000 U/min, 20 °C für die Dauer von 70 min wurde der Becherinhalt von 5 ml mit Hilfe einer speziellen Abtropfvorrichtung in 35 Fraktionen aufgetrennt; die einzelnen Fraktionen wurden auf Filterpapierscheiben (FN 7, VEB Spezialpapierfabrik Niederschlag) gesammelt. Nach dem Trocknen wurden die Papierscheiben 3mal mit eiskalter Trichloressigsäure behandelt und anschließend mit Äthanol und Aceton gewaschen. Die Radioaktivität ist mit Hilfe des Tricarb Liquid Scintillation Counter (Packard–Chicago) im Dioxan-System bestimmt worden.

Die spektroskopischen Messungen erfolgten mit dem UV-VIS-Spektrophotometer „Specord“ (VEB Carl Zeiss Jena).

Die photodynamische Inaktivierungskinetik von *P. mirabilis* bei Violamycin BI-Anwesenheit unterscheidet sich von der des Methylenblau. Beim Zusatz von Violamycin BI (Endkonzentrationen $1 \cdot 10^{-4}$ und $1 \cdot 10^{-5}$ M) erhaltene Inaktivierungskurven sind deshalb vergleichsweise mit der Inaktivierungskurve des photodynamisch sehr wirksamen Methylenblaus ($1{,}7 \cdot 10^{-6}$ M) in Abbildung 1 darstellt. Für vergleichende Wirkungsbetrachtungen ist bei Berücksichtigung der molaren Extinktionskoeffizienten, der unterschiedlichen Lage der Absorptionsbanden von MB und Violamycin und der unterschiedlichen Lichtemission der Wolframlampe bei verschiedenen Wellenlängen unter Vorbehalt die bei 10^{-4} M Violamycin erhaltene Inaktivierungskurve mit der für Methylenblau ($1{,}7 \cdot 10^{-6}$ M) angegebenen Kurve vergleichbar.

Orientierende Versuche mit Daunomycin ergaben, daß in ähnlichen Konzentrationsbereichen eine dem Violamycin BI vergleichbare photodynamische Wirkung erzielt wird. Die Inaktivierungskinetik bei Daunomycineinsatz entspricht allerdings mehr der bei der Methylenblau beobachteten (vgl. auch Verini *et al.* 1968).

Die bei Verwendung von Methylenblau zu beobachtenden Kurvenverläufe entsprechen in der Form denen, die allgemein mit photodynamisch wirksamen Farbstoffen erhalten werden. Sie sind auch vergleichbar mit Überlebenskurven nach UV- und Röntgeninaktivierung. Im Gegensatz dazu beobachtete man bei Anwesenheit von Violamycin BI zunächst eine schnelle Inaktivierung, an die sich eine Periode der weiteren, aber geringen Abnahme überlebender Bakterien anschließt. Spektrophotometrische Messungen von belichteten und belüfteten Violamycin BI-Lösungen zeigen, daß durch Belichtung die Substanz abgebaut wird, die Höhe der Bande erniedrigt sich beträchtlich. Dies kann die Ursache für die an sich ungewöhnliche Verringerung der photodynamischen Inaktivierungsgeschwindigkeit sein. Es erscheint auch nicht ausgeschlossen, daß durch die Belichtung Violamycin BI irrever-

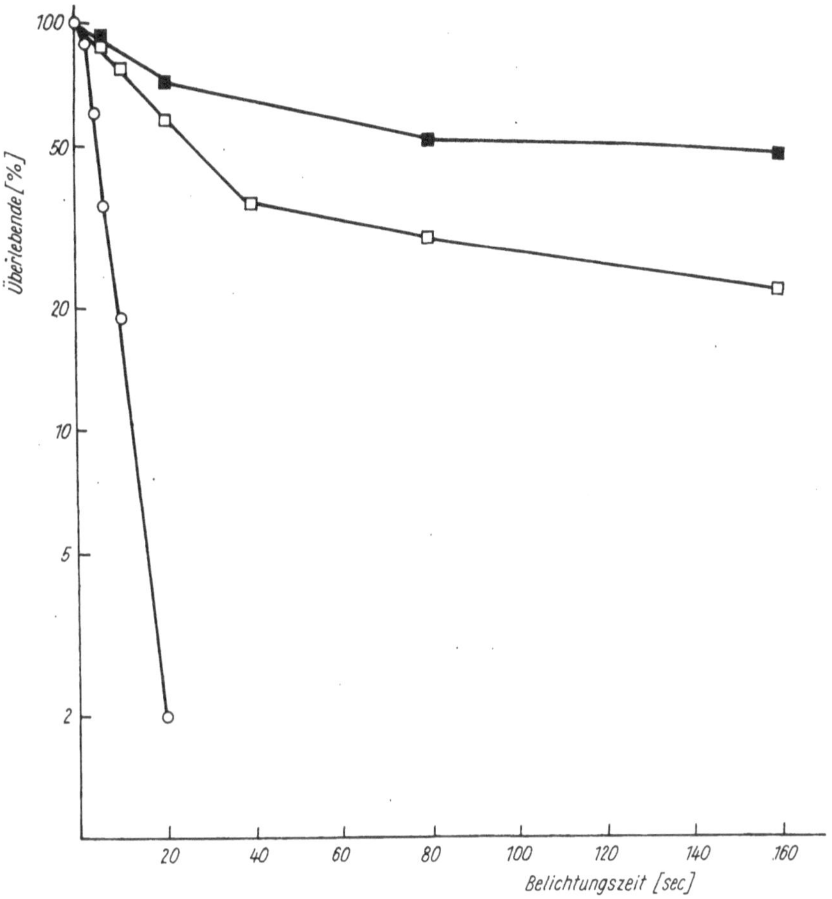

Abb. 1. Photodynamische Inaktivierungskinetik von *Proteus mirabilis* VI PG 273 mit Violamycin BI (□ = $1 \cdot 10^{-4}$ M, ■ = $1 \cdot 10^{-5}$ M und Methylenblau ○ = $1{,}7 \cdot 10^{-6}$ M)

sibel reduziert wird und dadurch für weitere Redox-Zyklen (Berg u. Jacob 1964) nicht mehr zur Verfügung steht.

Zur Überprüfung des Violamycin BI-Einflusses auf die Reparatur photodynamisch induzierter DNA-Schäden wurde das Anthracyclin-Antibioticum in Konzentrationen getestet, bei denen eine Dunkelwirkung auf nicht geschädigte Zellen ausbleibt. Wie aus Abbildung 2 mit der Legende ersichtlich, bewirkt Violamycin BI in einer Konzentration von $1 \cdot 10^{-4}$ M nur noch einen ganz geringen toxischen Effekt auf photodynamisch geschädigte Zellen von *P. mirabilis*, die Reparatur der DNA wird nicht beeinflußt. Bei einer Konzentration $5 \cdot 10^{-5}$ M ist keine toxische Wirkung und erwartungsgemäß auch kein Einfluß auf die Reparatur vorhanden. Im Gegensatz zu den bei *E. coli* K12 AB 1157 erhobenen Befunden, wonach Violamycin BI bei Konzentrationen von $5 \cdot 10^{-5}$ M die Reparatur von UV-induzierten DNS-Schäden deutlich stört (Störl *et al.* 1977), sind unter den beschriebenen Bedingungen bei *P. mirabilis* in Gegenwart von Violamycin BI keine störenden Einflüsse auf Reparaturprozesse photodynamisch induzierter DNA-Schäden zu beobachten.

Wird Daunomycin auf die Hemmwirkung der Reparatur photodynamischer Schäden getestet, so zeigt die Konzentration $1 \cdot 10^{-4}$ M eine deutliche Inhibition

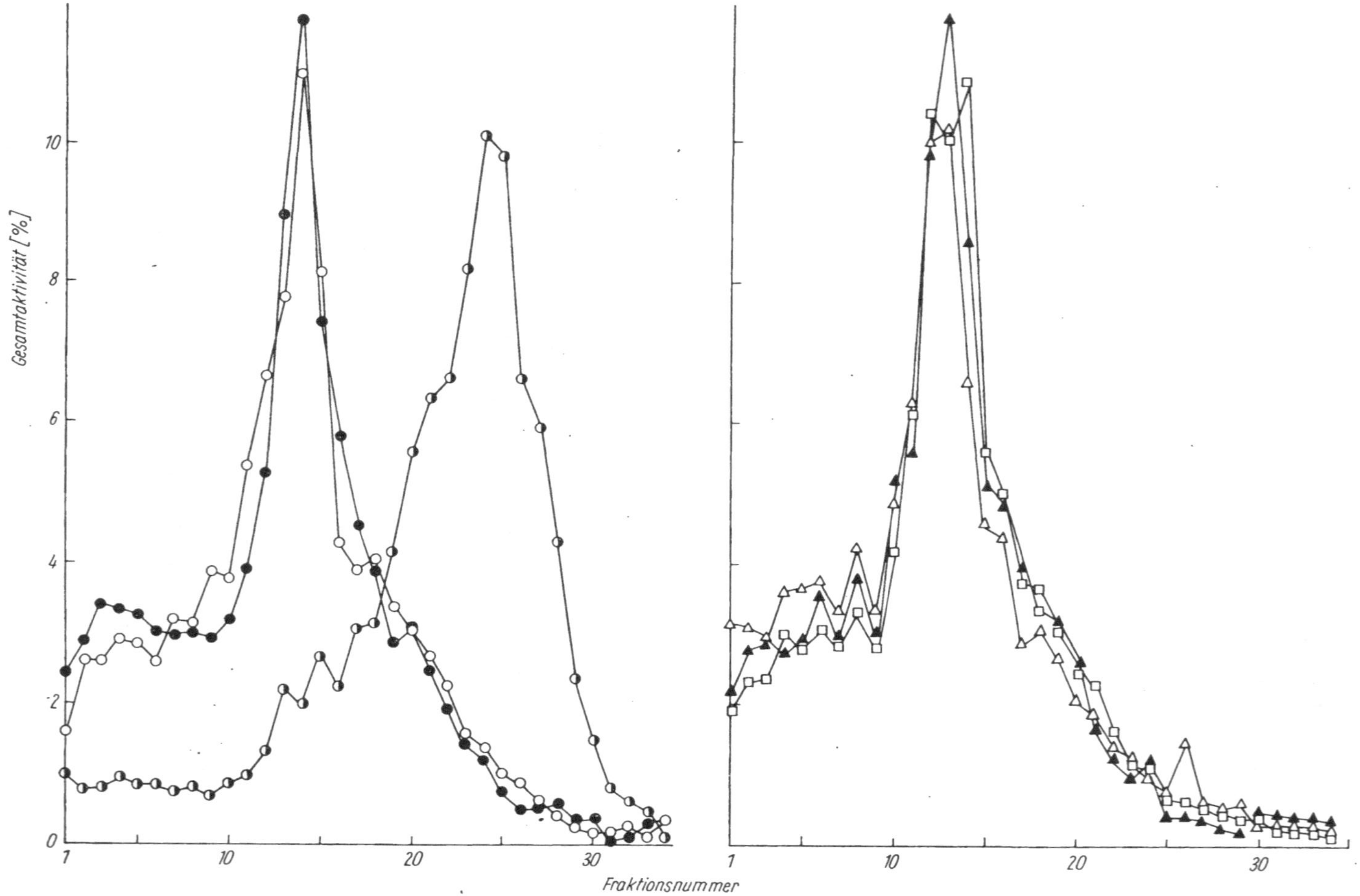

Abb. 2. Sedimentationsprofile von radioaktiv markierter DNS. Sedimentation von rechts nach links. ○ Kontrolle (t_0) 100% Bakterien, ◑ inaktivierte Probe ($t_{8,5}$ sec) 89% Überlebende, ● Reparatur nach Inaktivierung 20 min in Nährmedium 123% Überlebende, △ wie 3, mit $1{,}0 \cdot 10^{-4}$ M Violamycin BI 60% Überlebende, □ wie 3, mit $5{,}0 \cdot 10^{-5}$ M Violamycin BI 123% Überlebende, ▲ wie 3, mit $2{,}5 \cdot 10^{-5}$ M Violamycin BI 125% Überlebende

der Reparatur, aber es ist auch ein beträchtlicher toxischer Einfluß auf die photodynamisch behandelten Bakterien vorhanden. Bei einer Konzentration von $2{,}5 \cdot 10^{-5}$ M ist keine Reparaturhemmung mehr zu beobachten, der toxische Effekt ist sehr gering.

Mit Adriamycin sind die erhaltenen Ergebnisse ähnlich. Bei $2 \cdot 10^{-5}$ M wird eine deutliche toxische Wirkung beobachtet, die Hemmung der Reparatur ist gering. Bei $1 \cdot 10^{-5}$ M besteht zwar noch ein geringer Einfluß auf die Überlebenszahl, die Reparatur wird aber nicht mehr beeinflußt.

Den drei Antibiotica ist gemeinsam, daß sie Reparaturhemmeffekte nur zeigen, wenn sie noch eine beträchtliche toxische Wirkung auf die inaktivierten Bakterien ausüben. Werden Konzentrationen gewählt, die keine toxische Wirkung haben, so tritt keinerlei Beeinflussung der Reparaturprozesse nach photodynamischer Inaktivierung auf. Die Antibiotica Daunomycin und Adriamycin haben bei vergleichbaren Konzentrationen größere toxische Wirkungen als Violamycin BI.

Literatur

Berg, H. und Jacob, H.-E., 1964. Biologische Wirkungen photochemischer Wasserstoffperoxydbildung. 2. Intensivierung der photodynamischen Wirkung von Anthrachinon-2-sulfonsäure in Gegenwart von Glycerin und anderen Wasserstoffdonatoren. Photochem. Photobiol., **4**, 55–59.

Böhme, H., 1967. Genetic instability of an ultraviolet-sensitive mutant of *Proteus mirabilis*. Biochem. Biophysic. Res. Commun., **28**, 191–196.

Böhme, H., 1968. Absence of repair of photodynamically induced damage in two mutants of *Proteus mirabilis* with increased sensitivity to monofunctional alkylating agents. Mutation Res., **6**, 166–168.

DiMarco, A., Arcamone, F. and Zunino, F. 1975. Daunomycin (Daunorubicin) and adriamycin and structural analogues: biological activity and mechanism of action. In: J. W. Corocoran and F. E. Hahn (Editors), Antibiotics III. Mechanism of Action of Antimicrobial and Antitumor Agents, Springer-Verlag Berlin–Heidelberg–New York, pp. 101–128.

Fleck, W., Strauss, D., Koch, W. and Prauser, H., 1974. Violamycin, a new red-pigment antibiotic. Z. Allg. Mikrobiol., **14**, 551–558.

Fleck, W., Strauss, D., Koch, W. and Prauser, H., 1975. Biosynthesis and isolation of the new anthracycline antibiotics violamycin A, BI, BII and their aglycones (russ.). Antibiotiki (Moskau), **20**, 966–972.

Jacob, H.-E., 1975. DNA repair after photodynamic treatment of *Proteus mirabilis*. Photochem. Photobiol., **21**, 445–447.

Kersten, H. and Kersten, W., 1974. Inhibitors of nucleic acid synthesis. Biophysical and biochemical aspects. In: Molecular Biology, Biochemistry and Biophysics 18. Springer-Verlag Berlin-Heidelberg-New York, pp. 67–86.

Sanfilippo, A., Schiopacassi, G., Morvillo, E. and Ghione, M., 1968. Photodynamic action of daunomycin. I. Effect on bacteriophage T2 and bacteria. Giorn. Microbiol., **16**, 49–54.

Störl, K., Fleck, W. und Strauss, D., 1977. Einfluß von Violamycin BI auf den Ablauf der Excisionsreparatur in *Escherichia coli* K12 nach UV-Bestrahlung. Z. Allg. Mikrobiol., **17**, 459–464.

Verini, M. A., Casazza, A. M., Fioretti, A., Rhodenghi, F. and Ghione, M., 1968. Photodynamic action of daunomycin. II. Effect on animal viruses. Giorn. Microbiol., **16**, 55–64.

Zunino, F., Gambetta, R., DiMarco, A. and Zaccaro, A., 1975. A comparison of the effects of daunomycin and adriamycin on various DNA polymerases. Cancer Res., **35**, 754–760.

Anschrift: Dr. H.-E. Jacob
Zentralinstitut für Mikrobiologie und experimentelle Therapie der AdW
DDR 69 Jena, Beuthenbergstr. 11

Buchbesprechungen

Current Topics in Microbiology and Immunology, Vol. 74. III, 168 S., 22 Abb., 10 Tab. Berlin-Heidelberg-New York 1976: Springer-Verlag. DM 78,00.

Fünf Beiträge machen den Inhalt dieser bewährten Publikationsreihe aus, wovon drei besonders den Leser dieser Zeitschrift ansprechen werden. V. PIRROTTA faßt in seinem Artikel über den λ-Repressor und seine Aktion besonders die nach 1971 erschienenen Beiträge über den λ-Coliphagen zusammen. An der Isolierung des Repressorproteins und seiner elektronenmikroskopischen Darstellung hat die Arbeitsgruppe des Autors selbst einen bedeutsamen Anteil. Mit Hilfe der Bindung der Repressoren an die Operatoren konnten auch letztere isoliert und sichtbar gemacht werden. — K. GEIDER behandelt die molekularen Aspekte der DNA-Replikation in *Escherichia coli*-Systemen, wobei im besonderen auf die *in vivo*- und *in vitro*-DNA-Replikation eingegangen wird, auf RNA als Initiator der DNA-Replikation, auf die beteiligten Proteine und auf die Replikation isolierter DNA und gereinigter Proteine. — Teichosäuren der Bakterien haben in letzter Zeit das besondere Interesse der Mikrobiologen gefunden. Die seit 1960 berücksichtigten Publikationen stellen mit 171 Zitaten nur eine Auswahl dar und werden der Fülle des Materials wohl nicht vollauf gerecht (Beitrag von R. DZIARSKI: Teichoic acids). — Die beiden übrigen Beiträge des Bandes A. RIMON: Chemische und immunchemische Identität des Amyloids und P. A. SHARP und S. J. FLINT: Adenovirus Transcription, werden den entsprechenden Fachkollegen von Interesse sein. Auch dieser Band bedarf keiner besonderen Empfehlung. W. KÖHLER (Jena)

C. H. DICKINSON and T. F. PREECE (Editors), Microbiology of Aerial Plant Surfaces. IX, 669 S., 133 Abb., 79 Tab. London-New York-San Francisco 1976: Academic Press. £ 12.80.

Das Buch beinhaltet die 31 Vorträge eines Meetings, das im September 1975 in Leeds stattfand. Es ist eine Fundgrube für alle, die sich mit der mikrobiellen Ökologie im allgemeinen oder mit Fragen der Phytopathologie, der Symbiose, der Taxonomie und der Spezifität der mikrobiellen Besiedelung von Pflanzenoberflächen im speziellen beschäftigen. Der Fortschritt der Grundlagenforschung auf diesem insgesamt noch wenig entwickelten Gebiet wird deutlich. Ein wesentlicher Faktor ist dabei die zunehmende Anwendung der Rasterelektronenmikroskopie zur *in situ*-Darstellung der Mikroorganismen auf den Trägerpflanzen. U. TAUBENECK (Jena)

R. T. D. EMOND (ins Deutsche übertragen von W. D. GERMER), Farbatlas der Infektionskrankheiten. 384 S., 456 Abb., davon 407 mehrfarbig. Stuttgart-New York 1976: F. K. Schattauer Verlag. DM 56,00.

Dieser Farbatlas der Infektionskrankheiten ist ein Buch, das man sich seit langem wünschte. Obwohl die Vielfalt klinischer Bilder nur durch die eigene Erfahrung, durch die persönliche Konfrontation mit dem Patienten erkannt und erlernt werden kann, bietet dieser Atlas dem Arzt vielfache Möglichkeiten: Der Erfahrene kann sich mit den äußeren Bildern seltener Krankheiten vertraut machen, und der junge Mediziner erhält einen ersten Überblick über die typischen Formen der Infektionskrankheiten. Die bakteriellen Erkrankungen werden auf 135 Seiten abgehandelt, die Virusinfektionen nehmen den größeren Teil des Werkes ein. Zahlreiche Abbildungen wurden von Fachkollegen zur Verfügung gestellt. Das bot die Gewähr, daß die besten und typischsten Abbildungen ausgewählt werden konnten. Die Bilder, die dem Anliegen des Buches gemäß, den wesentlichen Teil des Werkes ausmachen, sind von hervorragender Qualität. Hier bleiben nur wenige Wünsche offen. Im Vordergrund stehen klinische Bilder, aber auch licht- und elektronenmikroskopische Abbildungen, sowie Farbphotographien von Bakterienkolonien kommen zu ihrem Recht. Die rechten Seiten sind den Abbildungen vorbehalten; ein erläuternder Text findet sich auf den gegenüberliegenden linken Seiten. Die Textdarstellung ist knapp und instruktiv. Problematisch bleibt nur die Postulierung eines Staphylokokken-Scharlachs. Selbst wenn das klinische Bild ähnlich ist, sollte man nicht von einem Scharlach sprechen. Die Bezeichnung *Corynebacterium diphtheriae mitis* bzw. *C. d. gravis* entspricht nicht den Nomenklaturregeln, es gibt nur jeweils einen Typ *mitis* bzw. *gravis* des *C. diphtheriae*. Besonders erfreulich ist auch der sehr geringe Preis, welcher der wünschenswerten weiten Verbreitung dieses Buches entgegenkommt. W. KÖHLER (Jena)

E. GEISSLER, E. LIBBERT, J. NITSCHMANN und G. THOMAS-PETERSEN (Herausgeber), Kleine Enzyklopädie. Das Leben. 896 S., über 300 meist mehrfarbige Abb. und Tab. Leipzig 1976: VEB Bibliographisches Institut. M 18,00.

In guter Ausstattung und zu einem erstaunlich niedrigen Preis legen Verlag, Herausgeber und ein Kollektiv von 35 Mitarbeitern eine Enzyklopädie der Lebenserscheinungen vor, geordnet nach den Stichworten, Kennzeichen und molekulare Grundlagen des Lebens — Zelle, — Gewebe-Organ-Organismus, — Individual-Entwicklung, — Evolution, — Organismen und Umwelt, — Angewandte Biologie, — Geschichte, philosophische Probleme der Biologie; das ganze ergänzt durch ein Sachregister. Für die Mikroorganismen wäre zu empfehlen, sie als dritte Gruppe von Pflanzen und Tieren abzutrennen, wie es schon PASTEUR und HAECKEL vorgeschlagen haben; Widersprüche bei dem Versuch, Mikroorganismen nach dem Schema von Pflanzen- und Tierzelle einzuordnen, ließen sich damit vermeiden. Prokaryota sind nicht stets einzellig (S. 136). Umgekehrt kann der Begriff Einzeller nicht auf Eukaryota beschränkt werden (S. 494), auch die Bakterien sind größtenteils Einzeller. Bakterien-Geißeln können je nach Spezies verschiedene Länge haben (S. 324). So ließen sich auch auf anderen Teilgebieten der Biologie Ungenauigkeiten finden, z. B. in der Frage des frei werdenden molekularen Wasserstoffs im anaeroben Stoffwechsel (S. 216).

Auch eine „kleine" Enzyklopädie verlangt eine in den Grundlagen umfassende Darstellung; dies Ziel dürfte erreicht sein. Was die Zuverlässigkeit der Daten betrifft, so ist diese Forderung schwerer zu erfüllen. Die Herausgeber mögen die Hinweise nicht als Kritik, sondern als Notizen für eine Neuauflage betrachten. W. SCHWARTZ (Braunschweig-Stöckheim)

M. R. HEINRICH (Editor), Extreme Environments. Mechanisms of Microbial Adaptation. XV, 362 S., 80 Abb., 53 Tab. New York-San Francisco-London 1976: Academic Press. $ 17.50.

Auf Meetings im Ames Research Center, Moffett Field, Californien wird alle zwei Jahre über Beziehungen zwischen Adaptation an extreme Lebensbedingungen und molekularen Zell-Mechanismen und über Möglichkeiten einer Anwendung auf andere Planeten, besonders auf den Mars und die dort vermuteten Lebensbedingungen, diskutiert. — Hier wird über das dritte Meeting 1974 berichtet, das M. ALEXANDER mit einem Vortrag über Beziehungen zwischen Selektion und Adaptation unter natürlichen und von Menschen produzierten extremen Bedingungen einleitet. Die Themen der 17 folgenden Vorträge behandeln in der Hauptsache die Zusammenhänge zwischen dem Leben bei hohen und niedrigen Temperaturen und dem Transfer von Informationen, der Aktivität von Enzymen und der Struktur und Funktion von Membranen. Zwei Vorträge berichten über mikrobiologische Studien in der Antarktis: Elektronenmikroskopie antarktischer Bodenbakterien (L. UYDESS u. W. V. VISHNIAC (†)); Lebensbedingungen für Mikroorganismen in den antarktischen Trockentälern (South Victoria Land) als Marsmodell (R. E. CAMERON, R. C. HONOUR, F. A. MORELLI).

Die Vorträge der ersten beiden Meetings sind nicht veröffentlicht worden. Ihre Themen-Gruppen werden im Anhang genannt. Das Buch ist dem Andenken von WOLF V. VISHNIAC gewidmet, der durch Arbeiten zu Fragen der extraterrestrischen Biologie bekannt geworden ist und bei Forschungsarbeiten in der Antarktis den Tod fand. W. SCHWARTZ (Braunschweig-Stöckheim)

G. M. HOFFMANN, F. NIENHAUS, F. SCHÖNBECK, H. C. WELTZIEN und H. WILBERT, Lehrbuch der Phytomedizin. 490 S., 101 Abb., 13 Tab. Berlin-Hamburg 1976: Verlag P. Parey. DM 88,00.

Seit etwa fünf Jahren erscheinen zunehmend Bücher über Pflanzenkrankheiten und Pflanzenschutz. Gegenüber der meist praxisbetonten Orientierung geht es den Autoren im vorliegenden Fall besonders darum, dem akademischen Nachwuchs den Zugang zu den Problemen der „wissenschaftlichen Phytomedizin" zu erleichtern und sie für diese Fachrichtung zu interessieren.

Im ersten der 6 Kapitel wird zunächst das Berufsbild des modernen Phytomediziners umrissen, und der Leser wird in die wirtschaftlichen Aspekte und gesetzlichen Maßnahmen im Pflanzenschutz eingeführt. Es folgt eine straff gegliederte Charakterisierung der einzelnen Schadorganismen höherer Pflanzen (Viren, Bakterien, Pilze, parasitische Blütenpflanzen und tierische Schädlinge) mit wesentlichen, stichwortartig zusammengefaßten Daten über Systematik, Entwicklung und verursachte Krankheiten. Kernstück des Buches sind zwei Kapitel, die Krankheitsentstehung, Verlauf und typische Erscheinungen behandeln. Eingehend werden die Wechselwirkungen zwischen Schadorganismus und Wirt und die bisher bekannten physiologischen Zusammenhänge im Verlauf der Infektion und Symptomausbildung behandelt. Leider wird dabei nicht immer deutlich gemacht, inwieweit es sich um Vermutungen oder gesicherte Ergebnisse handelt. Eine wünschenswerte Ergänzung bzw. Vereinfachung der ausführlich beschriebenen Befallsymptome wäre zweifellos mit der Wiedergabe photographierter Befallsbilder erreicht worden. Mit den wichtigsten Begriffen und Beziehungen aus der Populationsdynamik und Epidemiologie wird der Leser im 5. Kapitel vertraut gemacht. In zahlreichen Beispielen werden die eingeführten, gut definierten Begriffe erläutert; wesentliche Zusammenhänge werden in beispielhaften tabellarischen Übersichten festgehalten. Das abschließende Kapitel behandelt den Pflanzenschutz. Auch hier ist die Vielfalt der Methoden und der chemischen Bekämpfungsmittel, die dem Phytomediziner zur Auswahl stehen, durch konsequente Gliederung klar und übersichtlich dargestellt: Dem „klassischen" Pflanzenschutz folgt ein Ausblick auf das Gebiet des integrierten Pflanzenschutzes, das Zukunftsfeld des Phytomediziners. Insgesamt kann man den Versuch der Autoren, ein modernes Lehrbuch zum Thema „die Pflanze als Patient" zu verfassen, als geglückt bezeichnen.

R. Lieberei (Braunschweig)

Christa Legler (Herausgeber), Ausgewählte Methoden der Wasseruntersuchung, Band 1: Chemische, physikalisch-chemische, physikalische und elektrochemische Methoden. Lieferung 3 (Schluß). 126 S., 8 Abb., 10 Tab. Jena 1976: VEB Gustav Fischer Verlag. DM 16,00.

Zu Band 1 der „Ausgewählten Methoden" ist jetzt die dritte (letzte) Lieferung erschienen (Lieferung 1 (1971): Z. Allg. Mikrobiol., **13**, 734 (1973), Lieferung 2 (1973): Z. Allg. Mikrobiol., **15**, 389 (1975)). Sie enthält Beiträge zur Anwendung der Dünnschichtchromatographie, der spezifischen Leitfähigkeitsmessung, der Trübungsmessung, desgleichen zur Bestimmung des Salzgehaltes und zum Nachweis von Ozon, Cobalt, Nitrit, Sulfat und Sulfit im Wasser. Es folgen eine Spezialmethode zur Erfassung niedriger Werte des Chemischen Sauerstoffbedarfs (CSB-Cr), die Eliminierungsanalyse mit Al-Oxiden, Graphit und Polyamid zur Trennung organischer Stoffgemische im Wasser und schließlich der Nachweis von Formaldehyd, von chlororganischen und phosphororganischen Insekticiden im Wasser.

Adelheid Schwartz (Braunschweig)

J. F. Peberdy, A. H. Rose, H. J. Rogers and E. C. Cocking (Editors), Microbial and Plant Protoplasts. XII, 370 S., 39 Abb., 32 Tab. London-New York-San Francisco 1976: Academic Press. £ 7.80.

Protoplasten haben als zellwandlose Formen von Bakterien, Hefen, Pilzen und Pflanzen zunehmende Bedeutung bei der Untersuchung der cytoplasmatischen Membran, der Transportmechanismen und der Übertragung von genetischem Material erlangt. Im vorliegenden Buch sind 20 Beiträge zusammengefaßt, die auf dem im September 1975 in Nottingham stattgefundenen „4. Internationalen Symposium über Hefe- und andere Protoplasten" gehalten wurden. Die Artikel konzentrieren sich auf vier thematische Schwerpunkte: Isolierung und Eigenschaften der Protoplasten, Interaktion von Protoplasten mit exogener DNS, Bakteriophagen und Viren, Fusion von Protoplasten sowie Protoplastenreversion und Zellwandbiosynthese. Jedes dieser Hauptthemen wird in jeweils 1 oder 2 Beiträgen für Bakterien-, Hefe-, Pilz- und Pflanzenprotoplasten behandelt. Bei den Artikeln, die von führenden Spezialisten geschrieben sind, handelt es sich um Übersichten, in die noch eigene Ergebnisse eingearbeitet sind.

Die Gegenüberstellung des Wissens über die Protoplasten der vier Organismengruppen unterstreicht die vorliegenden Gemeinsamkeiten und Unterschiede. Während die Umwandlung in Protoplastenzellen und deren osmotische Stabilisierung sowohl bei Mikroorganismen als auch bei Pflanzen keine prinzipiellen Schwierigkeiten mehr bereitet, stehen die Untersuchungen zur Aufnahme von exogenem genetischen Material und dessen Interaktion mit der zelleigenen DNS der Protoplasten erst in den Anfängen. Als die z. Zt. wichtigsten Gebiete der Protoplastenforschung werden in vorliegendem Buch die Fusion sowie die Zellwandbiosynthese und Protoplastenreversion hervorgehoben. Diese Prozesse erlauben eine somatische Hybridisierung und sind von praktischer Bedeutung für die Pflanzenzüchtung und für die Übertragung genetischen Materials bei Mikroorganismen.

Das Buch bietet in erster Linie eine gute vergleichende und umfassende Übersicht über die wichtigsten Ergebnisse und Probleme der gegenwärtigen Protoplastenforschung. Es enthält darüber hinaus wertvolle Informationen für alle zellbiologisch und genetisch orientierten Mikrobiologen und Botaniker, die sich mit diesen Zellformen näher beschäftigen.

J. Gumpert (Jena)

E. Quagliariello, F. Palmieri and T. P. Singer (Editors), Horizons in Biochemistry and Biophysics, Vol. 2. 340 S., 54 Abb., 9 Tab. Reading, Mass. 1976: Addison-Wesley Pbl. Co. $ 15.00.

Diese neue, in jährlichen Abständen erscheinende Publikationsreihe enthält jeweils *ca.* 9 Übersichtsreferate zu wesentlichen Erkenntnissen und Entwicklungen auf wichtigen biowissenschaftlichen Spezialgebieten. Anliegen des namhaften Herausgeberkollegiums ist es, einen möglichst breiten Leserkreis über das eigene Arbeitsgebiet hinaus für Probleme anderer Forschungsrichtungen zu interessieren und so dem Trend zu wachsender Abgrenzung der einzelnen Disziplinen entgegenzuwirken.

Der nunmehr vorliegende zweite Band enthält wiederum in klarem und verständlichem Stil gehaltene Beiträge international anerkannter Autoren, wobei der Schwerpunkt mit 5 Referaten aktuellen Fragen des zellulären Transports zugeordnet wurde. So befassen sich H. L. KORNBERG und P. J. F. HENDERSON mit Vorteilen und Möglichkeiten des Einsatzes von Mutanten beim Studium des mikrobiellen Kohlenhydrattransports. A. FONYO, F. PALMIERI und E. QUAGLIARIELLO referieren über den Metabolittransport in Mitochondrien. Dabei wird insbesondere auf Nachweis, Funktion und Rolle verschiedener Carrier-Systeme unter Diskussion damit verbundener Probleme eingegangen. Der Transport von Zuckern und Aminosäuren in tierischen Zellen ist Gegenstand eines weiteren Beitrages (U. HOPFER). H. W. HELDT diskutiert den Mechanismus des Transfer von Substraten durch die Chloroplastenhülle, wobei insbesondere die Funktion der Translokationssysteme beim Austausch von Metaboliten, in der CO_2-Fixierung und in der Bildung von Polysacchariden Berücksichtigung finden. Ergebnisse von Untersuchungen zur Abhängigkeit des Ionentransports durch Lipid-Doppelschichten von elektrischen Potentialänderungen und dem Assoziationsgrad der Translokatormoleküle werden in einem anderen Artikel vorgestellt (P. MUELLER).

Die weiteren vier Beiträge sind dagegen recht unterschiedlichen Themen gewidmet. M. SEMERIVA und P. DESNUELLE erläutern am Beispiel der Pankreaslipase und Colipase Prinzipien der Biokatalyse in heterogener Phase. Hervorgehoben werden die Substrataktivierung durch Emulsionsbildung an der Grenzfläche, Struktur- und Funktionsbeziehungen sowie der Colipase-Effekt als Beispiel einer Protein-Protein-Wechselwirkung in der Lipidphase. In einem Bericht von M. S. HERSHFIELD und J. E. SEEGMILLER über die Regulation des Purinstoffwechsels werden mögliche Ursachen einer Überproduktion von Harnsäure und damit des Auftretens der Gicht besprochen. Ein anderes Referat ist den Perspektiven der Anwendung optisch-spektroskopischer Methoden, insbesondere der Fluoreszenzspektroskopie, bei der Aufklärung der Struktur und Funktion von Biomolekülen gewidmet (G. WEBER). Ein Modell für die 'horizontale' Genexpression im Lebenszyklus der Eukaryotenzelle wird von P. VOLPE vorgestellt. Die Aussagen stützen sich vor allem auf die Analyse des chronologischen Ablaufs der Makromolekülsynthese in synchronisierten *HeLa*-Zellkulturen.

U. GRÄFE (Jena)

N. R. ROSE and H. FRIEDMAN (Editors), Manual of Clinical Immunology. 932 S., 192 Abb., 151 Tab. Washington 1976: American Society for Microbiology. $ 20.00.

Dieses Buch ist das Gegenstück zum „Manual of Clinical Microbiology", das gleichfalls von der American Society for Microbiology herausgegeben wurde und deren sowie der American Association of Immunologists Empfehlungen für die immunologischen Tests enthält. In 124 Kapiteln werden von 185 Autoren sämtliche für die klinische Immunologie bedeutsamen Methoden abgehandelt: Humorale und zelluläre Komponenten der Immunantwort; Radioimmunassays; Bakterien-, Pilz-, Parasiten-; Virus-, Rickettsien- und Chlamydienserologie; Immunhämatologie; Test bei Allergien und Immunmangelkrankheiten; Autoimmunerkrankungen; Tumor- und Transplantationsimmunologie. Die angegebenen Methoden können auch vom erfahrenen technischen Assistenten direkt nachgearbeitet werden.

Erstaunlich und erfreulich ist die Kürze der Darstellung, die nicht nur technische Einzelheiten enthält, sondern auch Bewertungskriterien, Fehlerquellen, einige grundlegende Informationen, klinische Bedeutung usw. Den meisten Beiträgen ist auch ein Bezugsquellenverzeichnis der Reagentien beigefügt — fast ausschließlich Firmen der USA. Die serologische Differenzierung der Mikroorganismen wird in diesem Werk naturgemäß nicht berücksichtigt. Über einige Tests mag man geteilter Meinung sein — so z. B. die ausschließliche Empfehlung von Latextests für den Rheumafaktorennachweis — aber insgesamt betrachtet ist es ein praxisnahes, durchaus empfehlenswertes Laborhandbuch.

W. KÖHLER (Jena)

R. E. STRANGE, Microbial Response to Mild Stress (Reihe: Patterns of Progress/Microbiology Series, No. 6). 83 S., 14 Abb. Shildon, Co. Durham 1976: Meadowfield Press Ltd. $ 8.40.

Der Autor diskutiert in diesem Heft der Patterns of Progress-Reihe die Prozesse, die bei Bakterien ablaufen, wenn sie Einflüssen ausgesetzt sind, die nicht zum sofortigen Absterben, wohl aber zu deutlichen, in unterschiedlichem Maße wieder ausgleichbaren Schäden der Zellen führen. Behandelt werden Hunger, Wärme, Kälteschock, osmotischer Schock und Aerosolisierung. Neben dem Versuch, die Phänomene mit den neueren Erkenntnissen über den Ablauf des endogenen Stoffwechsels, die Induktion von Permeabilitätsveränderungen *etc.* zu erklären, werden auch Hinweise auf die praktische Bedeutung gegeben, die das Verständnis der besprochenen Vorgänge hat.

Probleme der Überwachung von Sterilisationsprozessen, der Aufbewahrung von Bakteriensuspensionen bei Kühlschranktemperaturen, der Deutung der Ergebnisse von Vitalfärbungen *etc.* werden in diesem Zusammenhang genannt. — Eine anregende Lektüre für jeden Mikrobiologen.

U. TAUBENECK (Jena)

D. J. WEBER and W. M. HESS (Editors), The Fungal Spore. Form and Function. 895 S., 282 Abb., 38 Tab. New York-London-Sydney-Toronto 1976: John Wiley & Sons. £ 18.15.

Die Brigham Young University, Provo, Utah, veranstaltete 1974 das II. Internationale Pilzsporensymposium. Das vorliegende Buch enthält die erweiterten und mit Literaturangaben versehenen Vorträge.

Kapitel 1: „Dormancy and Activation" umfaßt drei Teilgebiete, die sich mit der Ruhespore (HAWKER u. MADELIN, 70 S.), der Keimungs-Selbsthemmung (MACKO *et al.*, 28 S.) und den Aktivatoren der Sporenkeimung befassen (SUSSMAN, 40 S.). Kapitel 2 behandelt den Kohlenhydrat-, Lipid-, Protein-, Nucleinsäure-Stoffwechsel und die sich verändernden Organellen bei der Sporenkeimung sowie die Zellwandzusammensetzung bei der Keimschlauchbildung. Die Artikel stammen von GOTTLIEB, REISENER, LOVETT, VAN ETTEN *et al.*, SMITH *et al.* und AKAI *et al.* und umfassen insgesamt 272 Seiten. Kapitel 3 befaßt sich mit Form und Funktion der Pilzsporen. Abgehandelt werden Probleme der „resistenten Strukturen" bei *Myxomycetes* (das bezieht sich auf Sporenwand sowie alle festen Begrenzungen der Zellmasse). Weiterhin werden Belange bei Sporen von *Chytridiomycetes, Zygomycetes, Oomycetes* und *Basidiomycetes* (hier hauptsächlich Rostpilze) besprochen. Die Sporen der *Ascomycetes* wurden bezüglich ihrer kontrollierten Differenzierung bearbeitet, und ein Teilgebiet beschäftigt sich mit Conidien und ihrer Entwicklung. Die Vorträge stammen von ALDRICH/BLACKWELL, HOHL, CANTINO/MILLS, GROVE, BARTNICKI-GARCIA/HEMMES, HESS/WEBER, TURIAN und MANGENOT/REISINGER und umfassen 376 Seiten. Kapitel 4, als workshop bezeichnet, stammt von ELSIK und gibt einen Überblick (14 Seiten) über die fossilen Pilzsporen. Kapitel 5 besteht aus den aufgezeichneten Diskussionsbeiträgen der Teilnehmer anläßlich einer Aussprache zu den Trends in künftiger Pilzsporenforschung. Ein ausführlicher Index schließt das Buch ab.

Das Werk ist drucktechnisch (Schrift und Abbildungen) einwandfrei und enthält 282 elektronenmikroskopische Abbildungen. Darunter befinden sich sehr viele Aufnahmen, die mit dem Scanning-Mikroskop hergestellt wurden. Weitere Erläuterungen sind auf 43 Strichzeichnungen und in 38 Tabellen zu finden. Durch die Erfassung der Pilzsporenliteratur bis 1974 wurde die seit dem 1. Symposium (1965) entstandene Literaturlücke geschlossen.

R. TRÖGER (Jena)

G. WILDFÜHR und W. WILDFÜHR, Medizinische Mikrobiologie, Teil I und II. Ein Grundriß für Studenten. XV und XI, 564 S., 93 Abb., 23 Tab. Berlin 1975: VEB Verlag Volk und Gesundheit. M 28,00.

Eine medizinische Mikrobiologie im Taschenbuchformat, alle Daten auf die kürzeste Form gebracht, wichtiges durch Fettdruck hervorgehoben, so bieten sich die beiden Bändchen dem Leser an. Die Einteilung folgt dem üblichen Schema: Kurzer allgemeiner Teil mit einer ersten Einführung in mikrobiologische Grundlagen, Epidemiologie, Immunbiologie, Desinfektion auf 161 Seiten und der umfangreichere spezielle Teil (336 Seiten), der, ergänzt durch einfache schematisierte Zeichnungen, menschenpathogene Mikroorganismen, Viren, Protozoen, Helminthen behandelt.

W. SCHWARTZ (Braunschweig-Stöckheim)

Z. YOSHII, J. TOKUNAGA and J. TAWARA (Editors), Atlas of Scanning Electron Microscopy in Microbiology. VI, 233 S., 205 Abb., 2 Tab. Stuttgart-Tokyo 1976: Georg Thieme Verlag — Igaku Shoin Ltd. DM 160,00.

Der Atlas bringt in der Einleitung einen kurzen Überblick über die physikalischen Grundlagen der Rasterelektronenmikroskopie, die Instrumente und ihren Betrieb, die Präparationsmethoden und einige Probleme der Auswertungsmoglichkeiten. Für alle hier berührten Gebiete gibt es inzwischen umfassendere Darstellungen.

Der Hauptteil beinhaltet 178 Aufnahmen ganz verschiedener Objekte: Bakterien, Spirochäten, Pilze (unter denen man auch *Nocardia* und *Streptomyces* entdeckt), kultivierte Zellen und Viren, Protozoen, Bakterien auf inneren Körperoberflächen (Ösophagus, Magen). Die Auswahl ist völlig zufällig und basiert auf Abbildungen, die von den Autoren und ihrem engeren Kollegenkreis unter ganz verschiedenen Gesichtspunkten produziert wurden. Die Relationen entsprechen in keiner Weise denen, die man von einem repräsentativen Querschnitt durch die Mikrobiologie erwarten sollte. Ein großer Teil der Abbildungen ist eindrucksvoll und die Wiedergabe insgesamt ausgezeichnet. Trotzdem bleibt der Atlas hinter den Erwartungen, die der Titel verspricht, weit zurück: Der hohe Preis wird für ein Werk verlangt, das nicht das allgemeine Interesse der Mikrobiologen beanspruchen kann. U. TAUBENECK (Jena)

J. M. YUHAS, R. W. TENNANT and J. D. REGAN (Editors), Biology of Radiation Carcinogenesis. 347 S., 111 Abb., 46 Tab. New York 1976: Raven Press. $ 25.00.

Der vorliegende Berichtsband enthält die 18 Vorträge eines internationalen Symposiums gleichen Titels, das im April 1975 in Gatlinburg (Tennessee) stattfand. Er dokumentiert nicht nur überzeugend das bis jetzt vorliegende empirische Material über die durch Röntgen-, γ- und Neutronenstrahlung induzierte Carcinogenese bei Mensch und Tier, sondern versucht auch eine Analyse der damit verbundenen Mechanismen auf zellulärer bzw. molekularer Ebene. Nach einer knappen Einleitung von VAN BEKKUM wird im ersten Kapitel der Einfluß physikalischer und biologischer Faktoren (Alter, Spezies, Geschlecht und genetische Prädisposition) auf die durch Strahlung verursachte maligne Transformation diskutiert und anschließend die Wechselwirkung derselben mit den Makromolekülen der Zelle beschrieben. Die letzten beiden Kapitel sind den experimentellen Ergebnissen von viralen und zellulären *in vitro*-Untersuchungen vorbehalten. Diese Einteilung erscheint zwangsläufig als etwas willkürlich, da die einzelnen Autoren meist den gleichzeitigen Einfluß von mehreren Faktoren auf die Carcinogenese untersuchen. Indem auch chemische und virologische Faktoren mit in die Betrachtung einbezogen werden, ergibt sich eine allgemeinere und umfassendere Darstellung dieses Spezialgebietes.

Das mit zahlreichen Abbildungen und Tabellen reichlich ausgestattete Buch vermittelt den derzeitigen Erkenntnisstand und „... perhaps more importantly what we do not know". Es ist dem Onkologen ebenso zu empfehlen, wie dem auf diesem Gebiet tätigen Biologen, Biochemiker, Virologen und Physiker. Der gute Gesamteindruck des Buches wird noch unterstrichen durch das meist sehr umfangreiche Literaturverzeichnis am Ende eines jeden Vortrages. K. AUGSTEN (Jena)

N. D. ZINDER (Editor), RNA Phages (Cold Spring Harbor Monograph Series). VIII, 428 S., 107 Abb., 37 Tab. Cold Spring Harbor 1975: Cold Spring Harbor Laboratory. $ 28.00.

In der Reihe der vom Cold Spring Harbor Laboratorium herausgegebenen Monografien erschien zuletzt der Band „RNA-Phagen".

In 14 Kapiteln werden Struktur, Replikation und Translation der RNA-Phagen behandelt, die zu den kleinsten bekannten Viren gehören. Der Band stellt eine repräsentative Auswahl aller auf dem Sektor der RNA-Phagen laufenden Arbeiten dar.

Im Kapitel 1 wird ein Überblick über die physikalischen Eigenschaften dieser Phagen und ihrer RNA-Moleküle gegeben, wobei gleichzeitig wichtige Informationen über grundlegende molekularbiologische Methoden mitgeteilt werden. Kapitel 2 befaßt sich mit genetischen Untersuchungen, die die Existenz von drei Komplementationsgruppen bei RNA-Phagen nachweisen. In dem Kapitel 3 werden die vom Phagen RNA-Molekül codierten Proteine behandelt, wobei Vergleiche zwischen Nucleotidsequenzen im Phagengenom und Aminosäuresequenzen angestellt werden. Eingeschlossen sind Ergebnisse über Initiations-Triplets und die Benutzung synonymer Codons. Die Physiologie der Phagenanheftung und des Infektionsprozesses wird in Kapitel 4 dargestellt. Vier weitere Kapitel befassen sich mit Problemen der Replikation und deren Kontrolle. Der Replikase-Komplex besteht aus virus- und wirtsspezifizierten Proteinen (Kapitel 5, 7, 8 und 9). Das Kapitel 6 ist der Morphogenese gewidmet. Dabei werden Probleme des assembling-Prozesses besonders ausführlich dargestellt.

Die Kapitel 10—12 behandeln die Translation und deren genetische Kontrolle, die sich selbst bei diesen einfach gebauten Viren als außerordentlich kompliziert erweist. Die chemische Struktur und biologische Aktivität der RNA des Phagen MS2 wird im Kapitel 13 ausführlich dargestellt, das außerdem die Sequenzanalyse des A-Proteins einschließlich zahlreicher methodischer Details enthält. Das letzte Kapitel behandelt schließlich RNA-Phagen anderer Bakterienspecies als *E. coli*.

Neben der Vermittlung wertvoller Erkenntnisse über eine besonders für Probleme der Sequenzanalyse wichtig gewordene Virusgruppe liegt der Wert des Buches besonders in der Darstellung modernster molekularbiologischer Methoden. Die Monografie von N. D. ZINDER kann deshalb allen genetisch und molekularbiologisch arbeitenden Gruppen wärmstens empfohlen werden.

S. KLAUS (Jena)